Random Networks for Communication

When is a network (almost) connected? How much information can it carry? How can you find a particular destination within the network? And how do you approach these questions – and others – when the network is random?

The analysis of communication networks requires a fascinating synthesis of random graph theory, stochastic geometry and percolation theory to provide models for both structure and information flow. This book is the first comprehensive introduction for graduate students and scientists to techniques and problems in the field of spatial random networks. The selection of material is driven by applications arising in engineering, and the treatment is both readable and mathematically rigorous. Though mainly concerned with information-flow-related questions motivated by wireless data networks, the models developed are also of interest in a broader context, ranging from engineering to social networks, biology, and physics.

MASSIMO FRANCESCHETTI is assistant professor of electrical and computer engineering at the University of California, San Diego. His work in communication system theory sits at the interface between networks, information theory, control, and electromagnetics.

RONALD MEESTER is professor of mathematics at the Vrije Universiteit Amsterdam. He has published broadly in percolation theory, spatial random processes, self-organised criticality, ergodic theory, and forensic statistics and is the author of *Continuum Percolation* (with Rahul Roy) and *A Natural Introduction to Probability Theory*.

Random Networks for Communication

From Statistical Physics to Information Systems

Massimo Franceschetti
and
Ronald Meester

CAMBRIDGE
UNIVERSITY PRESS

University Printing House, Cambridge CB2 8BS, United Kingdom

Cambridge University Press is part of the University of Cambridge.

It furthers the University's mission by disseminating knowledge in the pursuit of education, learning and research at the highest international levels of excellence.

www.cambridge.org
Information on this title: www.cambridge.org/9780521854429

© M. Franceschetti and R. Meester 2007

First published 2007

A catalogue record for this publication is available from the British Library

ISBN 978-0-521-85442-9 Hardback

Contents

Preface

What is this book about, and who is it written for? To start with the first question, this book introduces a subject placed at the interface between mathematics, physics, and information theory of systems. In doing so, it is not intended to be a comprehensive monograph and collect all the mathematical results available in the literature, but rather pursues the more ambitious goal of laying the foundations. We have tried to give emphasis to the relevant mathematical techniques that are the essential ingredients for anybody interested in the field of random networks. Dynamic coupling, renormalisation, ergodicity and deviations from the mean, correlation inequalities, Poisson approximation, as well as some other tricks and constructions that often arise in the proofs are not only applied, but also discussed with the objective of clarifying the philosophy behind their arguments. We have also tried to make available to a larger community the main mathematical results on random networks, and to place them into a new communication theory framework, trying not to sacrifice mathematical rigour. As a result, the choice of the topics was influenced by personal taste, by the willingness to keep the flow consistent, and by the desire to present a modern, communication-theoretic view of a topic that originated some fifty years ago and that has had an incredible impact in mathematics and statistical physics since then. Sometimes this has come at the price of sacrificing the presentation of results that either did not fit well in what we thought was the ideal flow of the book, or that could be obtained using the same basic ideas, but at the expense of highly technical complications. One important topic that the reader will find missing, for example, is a complete treatment of the classic Erdös–Rényi model of random graphs and of its more recent extensions, including preferential attachment models used to describe properties of the Internet. Indeed, we felt that these models, lacking a geometric component, did not fit well in our framework and the reader is referred to the recent account of Durrett (2007) for a rigorous treatment of preferential attachment models. Other omissions are certainly present, and hopefully similarly justified. We also refer to the monographs by Bollobás (2001), Bollobás and Riordan (2006), Grimmett (1999), Meester and Roy (1996), and Penrose (2003), for a compendium of additional mathematical results.

Let us now turn to the second question: what is our intended readership? In the first place, we hope to inspire people in electrical engineering, computer science, and physics to learn more about very relevant mathematics. It is worthwhile to learn these mathematics, as it provides valuable intuition and structure. We have noticed that there is a tendency to re-invent the wheel when it comes to the use of mathematics, and we thought it would be very useful to have a standard reference text. But also, we want to inspire mathematicians to learn more about the communication setting. It raises specific questions that are mathematically interesting, and deep. Such questions would be hard to think about without the context of communication networks.

In summary: the mathematics is not too abstract for engineers, and the applications are certainly not too mechanical for mathematicians. The authors being from both communities – engineering and mathematics – have enjoyed over the years an interesting and fruitful collaboration, and we are convinced that both communities can profit from this book. In a way, our main concern is the interaction between people at either side of the interface, who desire to *break on through to the other side*.

A final word about the prerequisites. We assume that the reader is familiar with basic probability theory, with the basic notions of graph theory and with basic calculus. When we need concepts that go beyond these basics, we will introduce and explain them. We believe the book is suitable, and we have used it, for a first-year graduate course in mathematics or electrical engineering.

We thank Patrick Thiran and the School of Computer and Communication Sciences of the École Politechnique Fédérale de Lausanne for hosting us during the Summer of 2005, while working on this book. Massimo Franceschetti is also grateful to the Department of Mathematics of the Vrije Universiteit Amsterdam for hosting him several times. We thank Misja Nuyens who read the entire manuscript and provided many useful comments. We are also grateful to Nikhil Karamchandani, Young-Han Kim, and Olivier Lévêque, who have also provided useful feedback on different portions of the manuscript. Massimo Franceschetti also thanks Olivier Dousse, a close research collaborator of several years.

List of notation

In the following, we collect some of the notation used throughout the book. Definitions are repeated within the text, in the specific context where they are used. Occasionally, in some local contexts, we introduce new notation and redefine terms to mean something different.

$\lvert \cdot \rvert$	Lebesgue measure
	Euclidean distance
	L_1 distance
	cardinality
$\lfloor \cdot \rfloor$	floor function, the argument is rounded down to the previous integer
$\lceil \cdot \rceil$	ceiling function, the argument is rounded up to the next integer
$\mathcal{A}$	an algorithm
	a region of the plane
a.a.s.	asymptotic almost surely
a.s.	almost surely
β	mean square constraint on the codeword symbols
B_n	box of side length $\sqrt{n}$
	box of side length n
$B_n^{\leftrightarrow}$	the event that there is a crossing path connecting the left side of B_n with its right side
$C(x)$	connected component containing the point x
C	connected component containing the origin
	channel capacity
$C(x, y)$	channel capacity between points x and y
	chemical distance between points x and y
C_n	sum of the information rates across a cut
$\partial(\cdot)$	inner boundary
$D(G)$	diameter of the graph G
$D(\mathcal{A})$	navigation length of the algorithm $\mathcal{A}$
d_{TV}	total variation distance
$E(\cdot)$	expectation
$g(x)$	connection function in a random connection model

$\overline{g}(|x|)$ connection function depending only on the Euclidian distance, i.e., $\overline{g} : \mathbb{R}^+ \to [0, 1]$ such that $\overline{g}(|x|) = g(x)$

G a graph

G_X generating function of random variable X

γ interference reduction factor in the SNIR model

$I(z)$ shot-noise process

$\tilde{I}(z)$ shifted shot-noise process

I indicator random variable

i.i.d.   independent, identically distributed

k_c critical value in nearest neighbour model

λ density of a Poisson process, or parameter of a Poisson distribution

λ_c critical density for boolean or random connection model

$\Lambda(x)$ density function of an inhomogeneous Poisson process

$\ell(x, y)$ attenuation function between points x and y

$l(|x - y|)$ attenuation function depending only on the Euclidian distance, i.e., $l : \mathbb{R}^+ \to \mathbb{R}^+$ such that $l(|x - y|) = \ell(x, y)$

N environmental noise

$N(A)$ number of pivotal edges for the event A

$N_\infty(B_n)$ number of Poisson points in the box B_n that are also part of the unbounded component on the whole plane

$N(n)$ number of paths of length n in the random grid starting at the origin

O origin point on the plane

P power of a signal, or just a probability measure

$Po(\lambda)$ Poisson random variable of parameter λ

p_c critical probability for undirected percolation

$\vec{p}_c$ critical probability for directed percolation

p_c^{site} critical probability for site percolation

p_c^{bond} critical probability for bond percolation

p_α critical probability for α-almost connectivity

$\psi(\cdot)$ probability that there exists an unbounded connected component

Q the event that there exists at most one unbounded connected component

r_α critical radius for α-almost connectivity in the boolean model

r_c critical radius for the boolean model

R rate of the information flow

$R(x, y)$ achievable information rate between x and y

$R(n)$ simultaneous achievable per-node rate in a box of area n

SNR   signal to noise ratio

SNIR   signal to noise plus interference ratio

T a tree

     a threshold value

$\theta(\cdot)$ percolation function, i.e., the probability that there exists an unbounded connected component at the origin

U the event that there exists an unbounded connected component

U_0	the event that there exists an unbounded connected component at the origin, when there is a Poisson point at the origin
W	channel bandwidth
	sum of indicator random variables
w.h.p.	with high probability
X	Poisson process
	a random variable
X_n	a sequence of random variables
X^m	a codeword of length m
$X(A)$	number of points of the Poisson process X falling in the set A
$X(e)$	uniform random variable in $[0, 1]$,
	where e is a random edge coupled with the outcome of X
$x \leftrightarrow y$	the event that there is a path connecting point x with point y
Z_n	nth generation in a branching process

1

Introduction

Random networks arise when nodes are randomly deployed on the plane and randomly connected to each other. Depending on the specific rules used to construct them, they create structures that can resemble what is observed in real natural, as well as in artificial, complex systems. Thus, they provide simple models that allow us to use probability theory as a tool to explain the observable behaviour of real systems and to formally study and predict phenomena that are not amenable to analysis with a deterministic approach. This often leads to useful design guidelines for the development and optimal operation of real systems.

Historically, random networks has been a field of study in mathematics and statistical physics, although many models were inspired by practical questions of engineering interest. One of the early mathematical models appeared in a series of papers starting in 1959 by the two Hungarian mathematicians Paul Erdös and Alfréd Rényi. They investigated what a 'typical' graph of n vertices and m edges looks like, by connecting nodes at random. They showed that many properties of these graphs are almost always predictable, as they suddenly arise with very high probability when the model parameters are chosen appropriately. This peculiar property generated much interest among mathematicians, and their papers marked the starting point of the field of random graph theory. The graphs they considered, however, were abstract mathematical objects and there was no notion of geometric position of vertices and edges.

Mathematical models inspired by more practical questions appeared around the same time and relied on some notion of geometric locality of the random network connections. In 1957, British engineer Simon Broadbent and mathematician John Hammersley published a paper introducing a simple discrete mathematical model of a random grid in which vertices are arranged on a square lattice, and edges between neighbouring vertices are added at random, by flipping a coin to decide on the presence of each edge. This simple model revealed extreme mathematical depth, and became one of the most studied mathematical objects in statistical physics.

Broadbent and Hammersley were inspired by the work they had done during World War II and their paper's motivation was the optimal design of filters in gas masks. The gas masks of the time used granules of activated charcoal, and the authors realised that proper functioning of the mask required careful operation between two extremes. At one extreme, the charcoal was highly permeable, air flowed easily through the cannister, but the wearer of the mask breathed insufficiently filtered air. At the other extreme, the charcoal pack was nearly impermeable, and while no poisonous gases got through, neither did sufficient air. The optimum was to have high charcoal surface area and tortuous paths for air flow, ensuring sufficient time and contact to absorb the toxin. They realised that this condition would be met in a critical operating regime, which would occur with very high probability just like Erdös and Rényi showed later for random graph properties, and they named the mathematical framework that they developed *percolation theory*, because the meandering paths reminded them of water trickling through a coffee percolator.

A few years later, in 1961, American communication engineer Edgar Gilbert, working at Bell Laboratories, generalised Broadbent and Hammersley's theory introducing a model of random planar networks in continuum space. He considered nodes randomly located in the plane and formed a random network by connecting pairs of nodes that are sufficiently close to each other. He was inspired by the possibility of providing long-range radio connection using a large number of short-range radio transmitters, and marked the birth of continuum percolation theory. Using this model, he formally proved the existence of a critical transmission range for the nodes, beyond which an infinite chain of connected transmitters forms and so long-distance communication is possible by successive relaying of messages along the chain. By contrast, below critical transmission range, any connected component of transmitters is bounded and it is impossible to communicate over large distances. Gilbert's ingenious proof, as we shall see, was based on the work of Broadbent and Hammersley, and on the theory of branching processes, which dated back to the nineteenth-century work of Sir Francis Galton and Reverend Henry William Watson on the survival of surnames in the British peerage.

Additional pioneering work on random networks appears to be the product of communication engineers. In 1956, American computer scientist Edward Moore and information theory's father Claude Shannon wrote two papers concerned with random electrical networks, which became classics in reliability theory and established some key inequalities, presented later in this book, which are important steps towards the celebrated threshold behaviours arising in percolation theory and random graphs.

As these early visionary works have been generalised by mathematicians, and statistical physicists have used these simple models to explain the behaviour of more complex natural systems, the field of random networks has flourished; its application to communication, however, has lagged behind. Today, however, there is great renewed interest in random networks for communication. Technological advances have made it plausible to envisage the development of massively large communication systems composed of small and relatively simple devices that can be randomly deployed and 'ad hoc' organise into a complex communication network using radio links. These networks can be used for human communication, as well as for sensing the environment and collecting and exchanging data for a variety of applications, such as environmental and habitat monitoring, industrial process control, security and surveillance, and structural health monitoring. The behaviour of these systems resembles that of disordered particle systems studied in statistical physics, and their large scale deployment allows us to appreciate in a real setting the phenomena predicted by the random models.

Various questions are of interest in this renewed context. The first and most basic one deals with connectivity, which expresses a global property of the system as a whole: can information be transferred through the network? In other words, does the network allow at least a large fraction of the nodes to be connected by paths of adjacent edges, or is it composed of a multitude of disconnected clusters? The second question naturally follows the first one: what is the network capacity in terms of sustainable information flow under different connectivity regimes? Finally, there are questions of more algorithmic flavour, asking about the form of the paths followed by the information flow and how these can be traversed in an efficient way. All of these issues are strongly related to each other and to the original 'classic' results on random networks, and we attempt here to give a unifying view.

We now want to spend a few words on the organisation of the book. It starts by introducing random network models on the infinite plane. This is useful to reveal *phase transitions* that can be best observed over an infinite domain. A phase transition occurs when a small variation of the local parameters of the model triggers a macroscopic change that is observed over large scales. Obviously, one also expects the behaviour that can be observed at the infinite scale to be a good indication of what happens when we consider finite models that grow larger and larger in size, and we shall see that this is indeed the case when considering scaling properties of finite networks. Hence, after discussing in Chapter 2 phase transitions in infinite networks, we spend some words in Chapter 3 on connectivity of finite networks, treating full connectivity and almost connectivity in various models. In order to deal with the information capacity questions in Chapter 5, we need more background on random networks on the infinite plane, and Chapter 4 provides all the necessary ingredients for this. Finally, Chapter 5 is devoted to studying the information capacity of a random network, applying the scaling limit approach of statistical physics in an information-theoretic setting, and Chapter 6 presents certain algorithmic aspects that arise in trying to find the best way to navigate through a random network.

The remainder of this chapter introduces different models of random networks and briefly discusses their applications. In the course of the book, results for more complex models often rely on similar ones that hold for simpler models, so the theory is built incrementally from the bottom up.

1.1 Discrete network models

1.1.1 The random tree

We start with the simplest structure. Let us consider a *tree T* composed of an infinite number of vertices, where each vertex has exactly $k > 0$ children, and draw each edge of the tree with probability $p > 0$, or delete it otherwise, independently of all other edges. We are then left with a random infinite subgraph of T, a finite realisation of which is depicted in Figure 1.1. If we fix a vertex $x_0 \in T$, we can ask how long is the line of descent rooted at x_0 in the resulting random network. Of course, we expect this to be on average longer as p approaches one. This question can also be phrased in more general

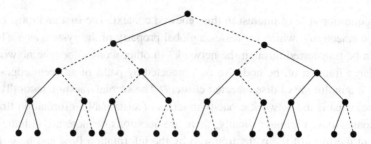

Fig. 1.1 A random tree $T(k, p)$, with $k = 2$, $p = 1/2$; deleted edges are represented by dashed lines.

terms. The distribution of the number of children at each node of the tree is called the *offspring distribution*, and in our example it has a Bernoulli distribution with parameters k and p. A natural way to obtain a random tree with arbitrary offspring distribution is by a so-called *branching process*. This has often been used to model the evolution of a population from generation to generation and it is described as follows.

Let Z_n be the number of members of the nth generation. Each member i of the nth generation gives birth to a random number of children, X_i, which are the members of the $(n + 1)$th generation. Assuming $Z_0 = 1$, the evolution of the Z_i can be represented by a random tree structure rooted at Z_0 and where

$$Z_{n+1} = X_1 + X_2 + \cdots + X_{Z_n}, \tag{1.1}$$

see Figure 1.2. Note that the X_i are random variables and we make the following assumptions,

 (i) the X_i are independent of each other,
 (ii) the X_i all have the same offspring distribution.

The process described above could in principle evolve forever, generating an infinite tree. One expects that if the offspring distribution guarantees that individuals have a sufficiently large number of children, then the population will grow indefinitely, with positive probability at least. We shall see that there is a critical value for the expected

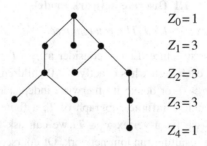

$Z_0 = 1$

$Z_1 = 3$

$Z_2 = 3$

$Z_3 = 3$

$Z_4 = 1$

Fig. 1.2   A random tree obtained by a branching process.

offspring that makes this possible and make a precise statement of this in the next chapter. Finally, note that the branching process reduces to our original example if we take the offspring distribution to be Bernoulli of parameters k and p.

1.1.2 The random grid

Another basic structure is the random *grid*. This is typically used in physics to model flows in porous media (referred to as percolation processes). Consider an infinite square lattice $\mathbb{Z}^2$ and draw each edge between nearest neigbours with probability p, or delete it otherwise, independently of all other edges. We are then left with a random infinite subgraph of $\mathbb{Z}^2$, see Figure 1.3 for a realisation of this on a finite domain. It is reasonable to expect that larger values of p will lead to the existence of larger connected components in such subgraphs, in some well-defined sense. There could in principle even be one or more infinite connected subgraphs when p is large enough, and we note that this is trivially the case when $p = 1$.

What we have described is usually referred to as a *bond percolation model* on the square lattice. Another similar random grid model is obtained by considering a *site percolation model*. In this case each box of the square lattice is occupied with probability p and empty otherwise, independently of all other boxes. The resulting random structure, depicted in Figure 1.4, also induces a random subgraph of $\mathbb{Z}^2$. This is obtained by calling boxes that share a side neighbours, and considering connected neighbouring boxes that are occupied. It is also interesting to note that if we take a tree instead of a grid as the underlying structure, then bond and site percolation can be viewed as the same process, since each bond can be uniquely identified with a site and vice versa.

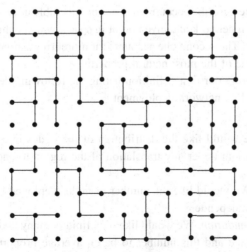

Fig. 1.3 The grid (bond percolation).

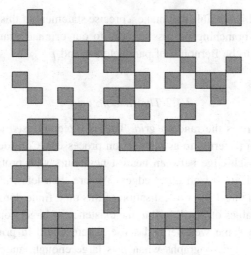

Fig. 1.4   The grid (site percolation).

1.2 Continuum network models

1.2.1 Poisson processes

Although stochastic, the models described above are developed from a predefined deterministic structure (tree and grid respectively). In continuum models this is no longer
the case as the positions of the nodes of the network themselves are random and are
formed by the realisation of a *point process* on the plane. This allows us to consider more
complex random structures that often more closely resemble real systems.

For our purposes, we can think of a point process as a random set of points on the
plane. Of course, one could think of a more formal mathematical definition, and we refer
to the book by Daley and Vere-Jones (1988) for this. We make use of two kinds of point
processes. The first one describes occurrences of unpredictable events, like the placement
of a node in the random network at a given point in space, which exhibit a certain amount
of statistical regularity. The second one accounts for more irregular network deployments,
while maintaining some of the most natural properties.

We start by motivating our first definition listing the following desirable features of a
somehow regular, random network deployment.

  (i)   *Stationarity*. We would like the distribution of the nodes in a given region of the
        plane to be invariant under any translation of the region to another location of the
        plane.
 (ii)   *Independence*. We would like the number of nodes deployed in disjoint regions of
        the plane to be independent.
(iii)   *Absence of accumulation*. We would like only finitely many nodes in every bounded
        region of the plane and this number to be on average proportional to the area of
        that region.

We now describe a way to construct a process that has all the features listed above and later give its formal definition. Consider first a square of side length one. Imagine we partition this square into n^2 identical subsquares of side length $1/n$ and assume that the probability p that a subsquare contains exactly one point is proportional to the area of the subsquare, so that for some $\lambda > 0$,

$$p = \frac{\lambda}{n^2}. \tag{1.2}$$

We assume that having two or more points in a subsquare is impossible. We also assume that points are placed independently of each other. Let us look at the probability that the (random) number of points N in the whole unit square is k. This number of points is given by the sum of n^2 independent random variables, each of which has a small probability λ/n^2 of being equal to one, and which are equal to zero otherwise. It is well known and not difficult to see that, as $n \to \infty$, this sum converges to the Poisson distribution of parameter λ, which is sometimes referred to as the *law of rare events*. Indeed,

$$
\begin{aligned}
\lim_{n \to \infty} P(N = k) &= \lim_{n \to \infty} \binom{n^2}{k} \left(\frac{\lambda}{n^2}\right)^k \left(1 - \frac{\lambda}{n^2}\right)^{n^2 - k} \\
&= \lim_{n \to \infty} \frac{n^2!}{k!(n^2 - k)!} \left(\frac{\lambda}{n^2}\right)^k \left(1 - \frac{\lambda}{n^2}\right)^{n^2} \left(1 - \frac{\lambda}{n^2}\right)^{-k} \\
&= \lim_{n \to \infty} \frac{\lambda^k}{k!} \left(1 - \frac{\lambda}{n^2}\right)^{n^2} \frac{n^2!}{n^{2k}(n^2 - k)!} \left(1 - \frac{\lambda}{n^2}\right)^{-k} \\
&= \frac{\lambda^k}{k!} e^{-\lambda}. \tag{1.3}
\end{aligned}
$$

The construction in the unit square clearly satisfies the three desired properties, and we now want to extend it to the whole plane. Consider two disjoint unit squares and look for the distribution of the number of points inside them. This is the sum of two independent Poisson random variables, and a simple exercise in basic probability shows that it is a Poisson random variable of parameter 2λ. This leads to the idea that in our point process on the plane, the number of points in any given region A should have a Poisson distribution of parameter $\lambda|A|$, where $|\cdot|$ denotes area. This intuition leads to the following definition.

Definition 1.2.1 (**Poisson process**) *A random set of points $X \subset \mathbb{R}^2$ is said to be a Poisson process of density $\lambda > 0$ on the plane if it satisfies the conditions*

(i) *For mutually disjoint domains of $\mathbb{R}^2$ $D_1, \ldots, D_k$, the random variables $X(D_1), \ldots, X(D_k)$ are mutually independent, where $X(D)$ denotes the random number of points of X inside domain D.*

(ii) *For any bounded domain $D \subset \mathbb{R}^2$ we have that for every $k \geq 0$*

$$P(X(D) = k) = e^{-\lambda|D|} \frac{(\lambda|D|)^k}{k!}. \tag{1.4}$$

Note that we have $E(X([0, 1]^2) = \lambda$ and the density of the process corresponds to the expected number of points of the process in the unit area. We also note that the definition does not say explicitly how to construct a Poisson process, because it does not say how the points are distributed on the plane, but only what the distribution of their number looks like.

However, a constructive procedure is suggested by the following observations. Let $B \subset A$ be bounded sets. By conditioning on the number of points inside A, and applying Definition 1.2.1, we have

$$P(X(B) = m \mid X(A) = m+k) = \frac{P(X(B) = m, \ X(A) = m+k)}{P(X(A) = m+k)}$$

$$= \frac{P(X(A \setminus B) = k, \ X(B) = m)}{P(X(A) = m+k)} = \frac{P(X(A \setminus B) = k) \, P(X(B) = m)}{P(X(A) = m+k)}$$

$$= \binom{m+k}{m} \left(\frac{|A| - |B|}{|A|} \right)^k \left(\frac{|B|}{|A|} \right)^m. \tag{1.5}$$

We recognise this expression as a binomial distribution with parameters $m+k$ and $|B|/|A|$. Hence, if we condition on the number of points in a region A to be $m+k$, then we can interpret the number of points that end up in $B \subset A$ as the number of successes in $m+k$ experiments with success probability $|B|/|A|$. This means that each of the $m+k$ points is randomly and uniformly distributed on A, and the positions of the different points are independent of each other.

Hence, to construct a Poisson point process in any bounded region A of the plane we should do the following: first draw a random number N of points from a Poisson distribution of parameter $\lambda|A|$, and then distribute these uniformly and independently over A.

Is it obvious now that this procedure indeed leads to a Poisson process, that is, a process that satisfies Definition 1.2.1? Strictly speaking, the answer is no: we described a necessary property of a Poisson process, but if we then find a process with this property, it is not yet clear that this property satisfies all the requirements of a Poisson process. More formally: the property is necessary but perhaps not sufficient. However, it turns out that it is in fact sufficient, and this can be seen by using a converse to (1.5) which we discuss next.

Suppose we have random variables N and $M_1, \ldots, M_r$ with the following properties: (i) N has a Poisson distribution with paramater μ, say; (ii) The conditional distribution of the vector $(M_1, \ldots, M_r)$ given $N = s$ is multinomial with parameters s and $p_1, \ldots, p_r$. We claim that under these conditions, $M_1, \ldots, M_r$ are mutually independent Poisson distributed random variables with parameters $\mu p_1, \ldots, \mu p_r$ respectively. To see this, we perform a short computation, where $m_1 + \cdots + m_r = s$,

$$P(M_1 = m_1, \ldots, M_r = m_r) = P(M_1 = m_1, \ldots, M_r = m_r | N = s) P(N = s)$$

$$= \frac{s!}{m_1! \cdots m_r!} p_1^{m_1} \cdots p_r^{m_r} e^{-\mu} \frac{\mu^s}{s!}$$

$$= \prod_{i=1}^{r} \frac{p_i^{m_i}}{m_i!} e^{-\mu p_i}, \tag{1.6}$$

proving our claim. The relevance of this is as follows: N represents the number of points in A in the construction above, and $M_1, \ldots, M_r$ represent the number of points ending up in regions $B_1, \ldots, B_r$ into which we have subdivided A. Since the properties of the construction are now translated into properties (i) and (ii) above, the conclusion is that the number of points in disjoint regions are mutually independent with the correct Poisson distribution. Hence we really have constructed a Poisson process on A.

Finally, we note that the independence property of the process also implies that if we condition on the event that there is a point at $x_0 \in \mathbb{R}^2$, apart from that point, the rest of the process is not affected by the conditioning event. This simple fact can be stated with arbitrarily high level of formality using *Palm calculus* and we again refer to the book of Daley and Vere-Jones (1988) for the technical details.

The definition of a Poisson point process can be generalised to the case when the density is not constant over the plane, but it is a function of the position over $\mathbb{R}^2$. This gives a non-stationary point process that is useful to describe non-homogeneous node deployments. We first describe a way to construct such a process from a standard Poisson point process and then give a formal definition. Let X be a Poisson point process with density λ on the plane, and let $g : \mathbb{R}^2 \to [0, 1]$. Consider a realisation of X and delete each point x with probability $1 - g(x)$, and leave it where it is with probability $g(x)$, independently of all other points of X. This procedure is called *thinning* and generates an inhomogeneous Poisson point process of density function $\lambda g(x)$. The formal definition follows.

Definition 1.2.2 **(Inhomogeneous Poisson process)** *A countable set of points $X \subset \mathbb{R}^2$ is said to be an inhomogeneous Poisson process on the plane with density function $\Lambda : \mathbb{R}^2 \to [0, \infty)$, if it satisfies the conditions*

(i) *For mutually disjoint domains of $\mathbb{R}^2$ $D_1, \ldots, D_k$, the random variables $X(D_1), \ldots, X(D_k)$ are mutually independent, where $X(D)$ denotes the random number of points inside domain D.*

(ii) *For any bounded domain $D \subset \mathbb{R}^2$ we have that for every $k \geq 0$*

$$P(X(D) = k) = e^{-\int_D \Lambda(x)dx} \frac{[\int_D \Lambda(x)dx]^k}{k!}. \tag{1.7}$$

In the case $\int_D \Lambda(x)dx = \infty$, this expression is interpreted as being equal to zero.

We claim that the thinning procedure that we decribed above leads to an inhomogeneous Poisson process with density function $\lambda g(x)$. To see this, we argue as follows.

We denote by $\tilde{X}$ the point process after the thinning procedure. The independence property is immediate from the construction, and the distribution of $\tilde{X}$ can be computed as follows:

$$P(\tilde{X}(A) = k) = \sum_{i=k}^{\infty} P(X(A) = i)P(\tilde{X}(A) = k \mid X(A) = i). \tag{1.8}$$

We have from (1.5) that given the event $\{X(A) = i\}$, the i points of X in A are uniformly distributed over A. Thus the conditional distribution of $\tilde{X}$ given $X(A) = k$ is just

$$P(\tilde{X}(A) = 1 \mid X(A) = 1) = |A|^{-1} \int_A g(x)dx, \qquad (1.9)$$

and more generally,

$$P(\tilde{X}(A) = k \mid X(A) = i) = \binom{i}{k} \left(|A|^{-1} \int_A g(x)dx \right)^k$$

$$\times \left(1 - |A|^{-1} \int_A g(x)dx \right)^{i-k}. \qquad (1.10)$$

Hence,

$$P(\tilde{X}(A) = k) = e^{-\lambda|A|} \frac{(\lambda \int_A g(x)dx)^k}{k!}$$

$$\times \sum_{i=k}^{\infty} \frac{(\lambda|A|[1 - |A|^{-1} \int_A g(x)dx])^{i-k}}{(i-k)!}$$

$$= e^{-\lambda|A|} \frac{(\lambda \int_A g(x)dx)^k}{k!} e^{\lambda|A|(1-|A|^{-1} \int_A g(x)dx)}$$

$$= \frac{(\lambda \int_A g(x)dx)^k}{k!} e^{-\lambda \int_A g(x)dx}. \qquad (1.11)$$

A few final remarks are appropriate. The definition of an inhomogeneous Poisson point process is more general than the described thinning procedure: since g is defined from $\mathbb{R}^2$ into $[0, \infty)$, it also allows accumulation points. Note also that $E(X([0, 1]^2) = \int_{[0,1]^2} \Lambda(x)dx$, which is the expected number of points of the process in the unit square. Finally, note that one can obtain a Poisson process from its inhomogeneous version by taking $\Lambda(x) \equiv \lambda$.

1.2.2 Nearest neighbour networks

We can now start to consider networks of more complex random structure. *Nearest neighbour networks* represent a natural mathematical construction that has been used, for example, to model multi-hop radio transmission, when a message is relayed between two points along a chain of successive transmissions between nearest neighbour stations.

Let X be a Poisson point process of unit density on the plane. We place edges between each point of X and its k nearest neighbours in Euclidean distance, where k is some chosen positive integer. The result is a random network, whose structure depends on the random location of the points and on the choice of k.

Note that the density of the Poisson process is just a scaling factor that does not play a role in the geometric properties of the graph. To see this, take a realisation of the Poisson process, and imagine scaling all lengths by a certain factor, say $1/2$. As shown

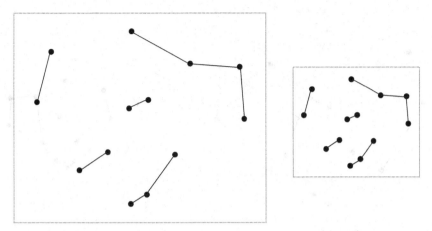

Fig. 1.5 Nearest neighbours model with $k = 1$. Changing the scale does not affect the connections between the nodes.

in Figure 1.5, this does not affect the connections between nearest neighbour nodes, while it increases the density of the points by a factor of 4. In other words, the model is *scale invariant*, and although the Euclidean distance between the nodes determines the connectivity of the network, it is the relative position of the nodes and not their absolute geometric coordinates that matters.

1.2.3 Poisson random connection networks

We now consider more geometric random networks. A *Poisson random connection model*, denoted by (X, λ, g), is given by a Poisson point process X of density $\lambda > 0$ on the plane, and a connection function $g(\cdot)$ from $\mathbb{R}^2$ into $[0, 1]$ satisfying the condition $0 < \int_{\mathbb{R}^2} g(x)dx < \infty$. Each pair of points $x, y \in X$ is connected by an edge with probability $g(x - y)$, independently of all other pairs, independently of X. We also assume that $g(x)$ depends only on the Euclidean norm $|x|$ and is non-increasing in the norm. That is,

$$g(x) \le g(y) \text{ whenever } |x| \ge |y|. \tag{1.12}$$

This gives a random network where, in contrast to the nearest neighbour model, the density λ of the Poisson process plays a key role, as densely packed nodes form very different structures than sparse nodes; see Figure 1.6.

The random connection model is quite general and has applications in different branches of science. In physics the random connection function may represent the probability of formation of bonds in particle systems; in epidemiology the probability that an infected herd at location x infects another herd at location y; in telecommunications the probability that two transmitters are non-shaded and can exchange messages; in biology the probability that two cells can sense each other.

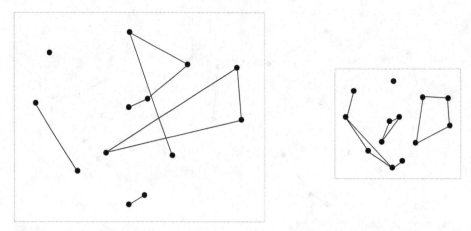

Fig. 1.6 A realisation of a random connection model. Changing the scale affects the connections between the nodes.

1.2.4 Boolean model networks

For a given $r > 0$, a special case of the random connection model is obtained when the connection function is of the boolean, zero-one type,

$$g(x) = \begin{cases} 1 \text{ if } |x| \leq 2r \\ 0 \text{ if } |x| > 2r. \end{cases} \tag{1.13}$$

This geometrically corresponds to placing discs of radius r at the Poisson points and considering connected components formed by clusters of overlapping discs; see Figure 1.7.

The *boolean model* can be adopted as a first-order approximation of communication by isotropic radiating signals. This applies to radio communication, and more generally to any communication where signals isotropically diffuse in a noisy environment. Some biological systems, for example, communicate by diffusing and sensing chemicals.

The main idea behind this application is to consider the circular geometries of the discs in Figure 1.7 as radiation patterns of signals transmitted by the Poisson points. Consider two points of the Poisson process and label them a transmitter x and a receiver y. The transmitter x radiates a signal with intensity proportional to the power P spent to generate the transmission. The signal propagates isotropically in the environment and is then received by y with intensity P times a loss factor $\ell(x, y) < 1$, due to isotropic dispersion and absorption in the environment. Furthermore, the reception mechanism is affected by noise, which means that y is able to detect the signal only if its intensity is sufficiently high compared to the environment noise $N > 0$. We conclude that x and y are able to establish a communication link if the signal to noise ratio (SNR) at the receiver is above a given threshold T. That is, if

$$SNR = \frac{P\ell(x, y)}{N} > T. \tag{1.14}$$

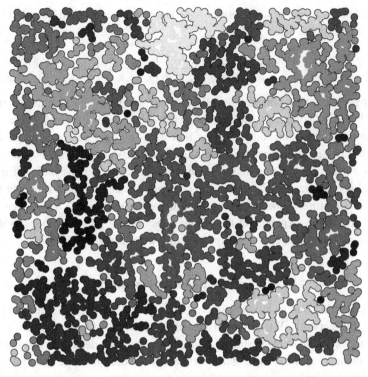

Fig. 1.7 Boolean model. Connected components of overlapping discs are drawn with the same grey level.

It is reasonable to assume the loss factor to be a decreasing function of the Euclidean distance between x and y. It follows that fixing the threshold T is equivalent to fixing the radius r of a boolean model where two nodes x and y are connected if they are sufficiently close to each other, that is, if discs of radius r centred at the nodes overlap.

1.2.5 Interference limited networks

The boolean model network represents the possibility of direct transmission between pairs of nodes at a given time, but the model does not account for the possible interference due to simultaneous transmission of other nodes in the network. In this case, all nodes can contribute to the amount of noise present at the receiver.

Consider two points of the Poisson process x_i and x_j, and assume x_i wants to communicate with x_j. At the same time, however, all other nodes x_k, $k \neq i, j$, also transmit an interfering signal that reaches x_j. We write the total interference term at x_j as $\gamma \sum_{k \neq i,j} P\ell(x_k, x_j)$, where $\gamma > 0$ is a weighting factor that depends on the technology adopted in the system to combat interference. Accordingly, we can modify the SNR model and obtain a *signal to noise plus interference ratio model* (SNIR model) by which x_j can receive messages from x_i if

$$SNIR = \frac{P\ell(x_i, x_j)}{N + \gamma \sum_{k \neq i,j} P\ell(x_k, x_j)} > T. \tag{1.15}$$

A random network is then obtained as follows. For each pair of Poisson points, the *SNIR* level at both ends is computed and an undirected edge between the two is drawn if this exceeds the threshold T in both cases. In this way, the presence of an edge indicates the possibility of direct bidirectional communication between the two nodes, while the presence of a path between two nodes in the graph indicates the possibility of multi-hop bidirectional communication. Note that the constructed random graph does not have the independence structure of the boolean model, because the presence of an edge between any pair of nodes now depends on the random positions of all other nodes in the plane that are causing interference, and not only on the two end-nodes of the link.

1.3 Information-theoretic networks

The models described up to now considered only the existence and Euclidean length of the links in the network. We have introduced the concept of communication between pairs of nodes saying that this is successful if the receiver can detect a transmitted signal, despite the attenuation due to the distance between the nodes, the noise due to the physical act of transmission, and the interference due to other simultaneous transmissions.

We are now interested in the *rate* at which communication can be performed in the random network. To precisely talk about this, however, we need some information theory concepts that will be introduced later in the book. Here, we want to give only an informal introduction to the information network model.

Let us consider a Poisson point process X of density $\lambda > 0$ on the plane, and a node $x_i \in X$ transmitting a signal of intensity P. Recall from our discussion of the boolean model that when a signal is radiated isotropically from x_i to x_j, it is in first approximation subject to a loss factor $\ell(x_i, x_j)$, due to diffusion and absorption in the environment. In the boolean model we have assumed that x_i and x_j are able to establish a communication link if the SNR at the receiver is above a certain threshold value. It is also reasonable to assume that this link can sustain a certain information flow, which is proportional to the SNR. In an information-theoretic network model, the relationship between the rate of the information flow and the SNR is non-linear and it is given by

$$R = \log\left(1 + \frac{P\ell(x_i, x_j)}{N}\right) \text{ bits per second.} \tag{1.16}$$

A physical justification for this formula will be given in Chapter 5. The formula shows that every pair of nodes in the network is connected by a link that can sustain a constant information rate, whose value depends on the loss factor between x_i and x_j, and hence by their random positions on the plane.

It is possible to modify (1.16) to account for the case of simultaneous communication of different nodes in the network. Let us assume for example that all nodes in the network are both transmitters and receivers, and each transmitter wishes to communicate to some receiver. How can we compute a possible rate between x_i and x_j in this case? A possible operation strategy is to treat all the interfering transmissions from other nodes on the same footing as random noise, so that an achievable rate is

$$R = \log\left(1 + \frac{P\ell(x_i, x_j)}{N + \sum_{k \neq i,j} P\ell(x_k, x_j)}\right) \text{ bits per second.} \tag{1.17}$$

By (1.17) we see that every pair of nodes can sustain a rate that depends not only on their relative position, but also on the (random) positions of all other nodes in the network. Note also that the infinite sum $\sum_{k \neq i,j} P\ell(x_k, x_j)$ in (1.17) can diverge, and in this case the rate between x_i and x_j goes to zero. If this happens, some strategy must be used to reduce the interference sum that appears in the denominator of (1.17) and obtain an achievable rate that is non-zero. For example, all nodes beside x_i and x_j could be kept silent, so that (1.17) reduces again to (1.16); or nodes could be used to *relay* information from x_i to x_j. Later, we are going to discuss some of these strategies and also show upper bounds on the achievable rate of communication that are *independent of any operation strategy*. Again, in order to talk about these we need some information theory background that we provide in Chapter 5. For now it is sufficient to keep in mind that the geometry of the information-theoretic network model is that of an infinite, fully connected graph, where vertices are Poisson points and the flow of information on each link is limited by the random spatial configuration of the nodes and by their transmission strategies.

1.4 Historical notes and further reading

The random tree network is known as a Galton–Watson branching process, after Watson and Galton (1874). The authors were concerned with the extinction of names in the British peerage. A large mathematical literature on branching processes is available today. Some of the main results are contained in the classic book of Harris (1963). The random grid network is known as a discrete percolation process. It was originally introduced in the classic paper by Broadbent and Hammersley (1957), and received much attention mostly in the statistical physics and mathematics literature. The extensive treatise of Grimmett (1999) is an excellent book for further reading. A formal introduction on point processes is given by the book of Daley and Vere-Jones (1988), while the book by Kingman (1992) focuses on Poisson processes alone. Nearest neighbour networks driven by a Poisson process were considered by Häggström and Meester (1996). The Poisson random connection model was considered by Penrose (1991), while the boolean model dates back to a paper by Gilbert (1961), which started the field of continuum percolation, the subject of the book by Meester and Roy (1996). Interference limited random networks, as described here, were considered by Dousse, Baccelli, and Thiran (2005); see also Dousse (2005). The question of achievable information rate in the limit of large, spatially random networks, was first considered by Gupta and Kumar (2000) in a slightly more restrictive communication model.

Exercises

1.1 Show that the sum of two independent Poisson random variables of parameter λ is still a Poisson random variable of parameter 2λ.

1.2 Generalise (1.5) as follows. Let $B_1, \ldots, B_n$ be disjoint sets, all contained in A. Compute $P(X(B_1) = m_1, \ldots, X(B_n) = m_n | X(A) = k + m_1 + \cdots + m_n)$. Do you recognise this distribution?

1.3 Show that the union of two independent inhomogeneous Poisson processes with density functions $\Lambda_1(x)$ and $\Lambda_2(x)$ respectively, is again an inhomogeneous Poisson process with density function $\Lambda_1(x) + \Lambda_2(x)$.

1.4 Show that in a Poisson process, with probability equal to one, no two pairs of points have the same distance to each other.

2

Phase transitions in infinite networks

One of the advantages of studying random network models on the infinite plane is that it is possible to observe sharp phase transitions. Informally, a phase transition is defined as a phenomenon by which a small change in the *local* parameters of a system results in an abrupt change of its *global* behaviour, which can be observed over an infinite domain. We shall see in subsequent chapters how these phenomena observed on the infinite plane are a useful indication of the behaviour in a finite domain. For now, however, we stick with the analysis on the infinite plane.

2.1 The random tree; infinite growth

We start by making a precise statement on the possibility that the branching process introduced in Chapter 1 grows forever. This is trivially true when the offspring distribution is such that $P(X_i \geq 1) = 1$, i.e., when each node in the tree has at least one child. However, it is perhaps less trivial that for generic offspring distribution it is still possible to have an infinite growth if and only if $E(X_i) = \mu > 1$.

Theorem 2.1.1 *When $\mu \leq 1$ the branching process does not grow forever with probability one, except when $P(X = 1) = 1$. When $\mu > 1$, the branching process grows forever with positive probability.*

The proof of Theorem 2.1.1 uses *generating functions*, so we start by saying a few words about these. Generating functions are a very convenient tool for all sorts of computations that would be difficult and tedious without them. These computations have to do with sums of random variables, expectations and variances. Generating functions are used only in this section, so readers not particularly interested in random trees can safely move on to the next section.

Definition 2.1.2 *Let X be a random variable taking values in $\mathbb{N}$. The* generating function *of X is defined as*

$$G_X(s) = E(s^X)$$

$$= \sum_{n=0}^{\infty} P(X = n)s^n, \tag{2.1}$$

for all $s \in \mathbb{R}$ for which this sum converges.

Clearly, $G_X(s)$ converges for at least all $s \in [0, 1]$. As an example, let X have a Poisson distribution with parameter λ. Then $G_X(s)$ can be computed as

$$G_X(s) = \sum_{n=0}^{\infty} e^{-\lambda} \frac{\lambda^n}{n!} s^n$$

$$= e^{-\lambda} \sum_{n=0}^{\infty} \frac{(\lambda s)^n}{n!}$$

$$= e^{-\lambda} e^{\lambda s}$$

$$= e^{\lambda(s-1)}. \tag{2.2}$$

The following result articulates the relation between expectations and generating functions.

Proposition 2.1.3 *Let X have generating function G. Then $E(X) = G'(1)$.*

Proof We have

$$G'(s) = \sum_{n=1}^{\infty} n s^{n-1} P(X = n)$$

$$\rightarrow \sum_{n=1}^{\infty} n P(X = n)$$

$$= E(X), \tag{2.3}$$

as s approaches one from below. □

Generating functions can also be very helpful in studying sums of random variables.

Proposition 2.1.4 *If X and Y are independent, then*

$$G_{X+Y}(s) = G_X(s) G_Y(s). \tag{2.4}$$

Proof Since X and Y are independent, so are s^X and s^Y. Hence

$$G_{X+Y}(s) = E(s^{X+Y})$$

$$= E(s^X s^Y)$$

$$= E(s^X) E(s^Y)$$

$$= G_X(s) G_Y(s). \tag{2.5}$$

□

This result clearly extends to any finite sum of random variables. The following lemma is very important for the study of branching processes and random trees since it deals with the sum of a *random* number of random variables.

Lemma 2.1.5 *Let $X_1, X_2, \ldots$ be a sequence of independent identically distributed random variables taking values in $\mathbb{N}$ and with common generating function G_X. Let N be a* random variable, independent of the X_i, also taking values in $\mathbb{N}$, with generating function G_N. Then the sum

$$S = X_1 + X_2 + \cdots + X_N \tag{2.6}$$

has generating function given by

$$G_S(s) = G_N(G_X(s)). \tag{2.7}$$

Note that if $P(N = n) = 1$, then $G_N(s) = s^n$ and $G_S(s) = (G_X(s))^n$, in agreement with Proposition 2.1.4.

Proof We write

$$G_S(s) = E(s^S)$$

$$= \sum_{n=0}^{\infty} E(s^S | N = n) P(N = n)$$

$$= \sum_{n=0}^{\infty} E(s^{X_1 + X_2 + \cdots + X_n}) P(N = n)$$

$$= \sum_{n=0}^{\infty} E(s^{X_1}) \cdots E(s^{X_n}) P(N = n)$$

$$= \sum_{n=0}^{\infty} (G_X(s))^n P(N = n)$$

$$= G_N(G_X(s)). \tag{2.8}$$

$\square$

Now we turn to branching processes proper. Recall from (1.1) that

$$Z_{n+1} = X_1 + X_2 + \cdots + X_{Z_n}, \tag{2.9}$$

where the X_i are independent random variables. Writing G_n for the generating function of Z_n, Lemma 2.1.5 and (2.9) together imply that

$$G_{n+1}(s) = G_n(G_1(s)), \tag{2.10}$$

and iteration of this formula implies that

$$G_n(s) = G_1(G_1(\cdots (G_1(s)) \cdots)), \tag{2.11}$$

the n-fold iteration of G_1. Note that G_1 is just the generating function of X_1, and we write $G = G_1$ from now on. In principle, (2.11) tells us everything about Z_n. For instance, we can now prove the following:

Proposition 2.1.6 *If $E(X_i) = \mu$, then $E(Z_n) = \mu^n$.*

Proof Differentiate $G_n(s) = G(G_{n-1}(s))$ at $s = 1$, and use Proposition 2.1.3 to find

$$E(Z_n) = \mu E(Z_{n-1}).\tag{2.12}$$

Now iterate this formula to obtain the result.   □

Hence, the expected number of members in a branching process grows or decays exponentially fast. If the expected number of children is larger than one, the expection grows to infinity, if it is smaller, it decays to zero, and this is consistent with Theorem 2.1.1. To actually prove this theorem we first prove the following.

Theorem 2.1.7 *The probability η that $Z_n = 0$ for some n is equal to the smallest non-negative root of the equation $G(s) = s$.*

Here is an example of Theorem 2.1.7 in action. Consider a random tree where each node has 0, 1 or 2 children (to be denoted by X) with probabilities given by $P(X = 0) = 1/8$, $P(X = 1) = 1/2$ and $P(X = 2) = 3/8$. The generating function G is now given by

$$G(s) = \frac{3}{8}s^2 + \frac{1}{2}s + \frac{1}{8}.\tag{2.13}$$

Solving $G(s) = s$ gives $s = 1/3$ and $s = 1$. The smallest non-negative solution is $s = 1/3$, and therefore the random tree is infinite with probability $2/3$.

Proof of Theorem 2.1.7 The probability η of ultimate extinction can be approximated by $\eta_n = P(Z_n = 0)$. Indeed, it is not hard to see that $\eta_n \to \eta$ as $n \to \infty$. We now write

$$\eta_n = P(Z_n = 0) = G_n(0) = G(G_{n-1}(0)) = G(\eta_{n-1}).\tag{2.14}$$

Now let $n \to \infty$ and use the fact that G is continuous to obtain

$$\eta = G(\eta).\tag{2.15}$$

This tells us that η is indeed a root of $G(s) = s$, but the claim is that it is the *smallest* non-negative root. To verify this, suppose that r is any non-negative root of the equation $G(s) = s$. Since G is non-decreasing on $[0, 1]$ we have

$$\eta_1 = G(0) \leq G(r) = r,\tag{2.16}$$

and

$$\eta_2 = G(\eta_1) \leq G(r) = r,\tag{2.17}$$

and so on, giving that $\eta_n \leq r$ for all n and hence $\eta \leq r$.   □

We are now ready for the proof of Theorem 2.1.1.

Proof of Theorem 2.1.1 According to Theorem 2.1.7, we need to look at the smallest non-negative root of the equation $G(s) = s$.

Suppose first that $\mu > 1$. Since

$$G'(1) = \mu,\tag{2.18}$$

we have that $G'(1) > 1$. Since $G(1) = 1$, this means that there is some $s' < 1$ for which $G(s') < s'$. Since $G(0) \geq 0$ and since G is continuous, there must be some point s'' between zero and s' with $G(s'') = s''$, which implies that the smallest non-negative solution of

$G(s) = s$ is strictly smaller than one. Hence the process survives forever with positive probability.

Next, consider the case in which $\mu \leq 1$. Note that

$$G'(s) = \sum_{n=1}^{\infty} ns^{n-1}P(X = n) > 0, \tag{2.19}$$

and

$$G''(s) = \sum_{n=2}^{\infty} n(n-1)s^{n-2}P(X = n) > 0, \tag{2.20}$$

where the strict inequalities come from the fact that we have excluded the case $P(X = 1) = 1$. This implies that G is strictly increasing and strictly convex. Hence if $G'(1) < 1$, then $G'(s) < 1$ for all $s \in [0, 1]$ and then it is easy to see that $G(s) > s$ for all $s < 1$, and therefore the smallest non-negative solution of $G(s) = s$ is $s = 1$, proving the result. □

Note that while for $\mu \leq 1$ the branching process does not grow to infinity with probability one, for $\mu > 1$ this is often possible only with some positive probability. However, we would like to see a sharper transition characterising this event. How can we make it certain? One problem of course is that we grow the process from one single point and if $P(X = 0) > 0$ this can always stop even at its first iteration with positive probability. One solution is then simply to restart the process. Recall the random tree that we introduced at the beginning of Section 1.1.1. Start with an infinite tree where each vertex has a fixed number of n children, and independently delete each edge with probability $1 - p$. We can see this as a branching process with Bernoulli offspring distribution with parameters n and p that starts $n - m$ new processes every time a node generates $m < k$ children. If the average number of offspring $np > 1$, then each time we start a new process, there is a positive probability that an infinite tree is generated. It follows that an infinite tree is generated, with probability one, after a finite number of trials. This probabilistic argument exploits the idea that, in order to obtain a sharp transition to probability one, one should not consider an event occurring at some specified vertex, but consider the same event occurring somewhere among an infinite collection of vertices. We see in the following how this type of argument can be adapted to cases where repetitions are not independent.

2.2 The random grid; discrete percolation

We now consider the random grid network with edge probability p (bond percolation). We define a connected component as a maximal set of vertices and edges such that for any two vertices x, y in the set, there exists an alternating sequence of distinct vertices and edges that starts with x and ends with y. In other words, x and y are in the same component if we can walk from one to the other over edges that are present. All the results we present also hold in the case of a random grid where each site is occupied independently with probability p (site percolation) and at the end of this section we shall see that very little is needed to accommodate the proofs. When the parameter is p, we write P_p for the probability measure involved.

A phase transition in the random grid occurs at a *critical value* $0 < p_c < 1$. Namely, when p exceeds p_c the random grid network contains a connected subgraph formed by an unbounded collection of vertices with probability one or, otherwise stated, almost surely (a.s.). In this case we say that the network *percolates*, or equivalently that the percolation model is *supercritical*. Conversely, when $p < p_c$ the random grid is a.s. composed of connected components of finite size, and we say that the model is *subcritical*. Next we prove these last statements.

Let $C(x)$ be the set of vertices connected to $x \in \mathbb{Z}^2$ and $|C(x)|$ its cardinality. We start by giving the following definition.

Definition 2.2.1 *The percolation probability $\theta(p)$ is the probability that the origin O (or any other vertex, for that matter) is contained in a connected component of an infinite number of vertices. That is, by denoting $C(O) \equiv C$,*

$$\theta(p) \equiv P_p(|C| = \infty) = 1 - \sum_{n=1}^{\infty} P_p(|C| = n). \tag{2.21}$$

We now want to study the function $\theta(p)$ in more detail. We start by noticing the trivial results: $\theta(0) = 0$ and $\theta(1) = 1$. Next, we show that:

Theorem 2.2.2 *When $0 < p_1 < p_2 < 1$, we have that $\theta(p_1) \leq \theta(p_2)$.*

This theorem is quite intuitive: increasing the edge probability p cannot decrease the chance of percolating. However, its simple derivation serves as an excellent illustration of a very powerful proof technique that we are going to use extensively throughout the book. This is called *coupling* and amounts to simultaneously constructing two realisations of two networks (one for p_1 and one for p_2) on the same probability space, and then noting that if there is a connected component in the first realisation, this is also true in the other. It then immediately follows that if the first model percolates, the other also does.

Proof of Theorem 2.2.2 We write

$$p_1 = p_2 \frac{p_1}{p_2}, \tag{2.22}$$

where $p_1/p_2 < 1$. Let G_p be the random grid of edge probability p. Consider a realisation of G_{p_2} and delete each edge independently from this realisation, with probability $(1 - p_1/p_2)$. The resulting graph can be effectively viewed as a realisation of G_{p_1} by virtue of (2.22). On the other hand, it is also clear that the latter realisation contains less edges that the original realisation of G_{p_2}. We conclude that if there is an infinite cluster in G_{p_1}, then there must also exist one in G_{p_2}, and this concludes the proof. □

The event $\{|C| = \infty\}$ is an example of a so-called *increasing* event. We are going to encounter similar events often in the book and it pays to give a formal definition here.

Definition 2.2.3 *An event A is increasing if adding an edge in any realisation of the random network where A occurs, leads to a configuration which is still in A. Similarly,* *a random variable is increasing if its value does not decrease by adding an edge to any realisation of the random network. Furthermore, an event is called decreasing if its complement is increasing, and a random variable is decreasing if its negative is increasing. An event (random variable) which is either increasing or decreasing is called monotone.*

It should be clear that the event $\{|C| = \infty\}$ is increasing: indeed, if the component of the origin is infinite, then this remains true if we add another edge. Another increasing event is the event that there exists an occupied path from vertex x to vertex y.

The following result generalises Theorem 2.2.2; the proof can be obtained along the same lines and it is left as an exercise.

Theorem 2.2.4 *For any increasing event A, the function*

$$p \to P_p(A) \tag{2.23}$$

is non-decreasing in $p \in [0, 1]$. For any decreasing event B, this function is non-increasing in p.

What we have learned so far is that $\theta(p)$ is a non-decreasing function of p that is zero in the origin and one at $p = 1$. The next theorem formally shows that the phase transition occurs at a non-trivial critical value p_c, that is, at a value strictly between zero and one.

Theorem 2.2.5 *There exists a $1/3 \le p_c \le 2/3$ such that $\theta(p) = 0$ for $p < p_c$ and $\theta(p) > 0$ for $p > p_c$.*

This theorem has the following basic corollary.

Corollary 2.2.6 *Let $\psi(p)$ be the probability of existence of an infinite connected component in the random grid. Then $\psi(p) = 0$ for $p < p_c$ and $\psi(p) = 1$ for $p > p_c$.*

The intuition behind this corollary is the probabilistic argument which can be informally stated as follows: if the probability of having an infinite connected component at some given vertex is positive, then the existence of an infinite component *somewhere* has probability one. The formal derivation, however, is slightly more complex because the event that there exists an infinite connected component at $x_1 \in \mathbb{Z}^2$ is not independent of the existence of an infinite component at $x_2 \in \mathbb{Z}^2$, and therefore we cannot simply write the probability of its occurrence at some $x \in \mathbb{Z}^2$ as

$$\lim_{n \to \infty} 1 - (1 - \theta(p))^n = 1, \tag{2.24}$$

as we have argued in the case of the random tree where realisations were independent.

Proof of Corollary 2.2.6 We first make use of Kolmogorov's zero-one law to show that the probability of an infinite cluster is either zero or one. This is a basic law to keep in mind when reasoning about events on the infinite plane. A formal statement of it can be found, for example, in the book by Grimmett and Stirzaker (1992). Here, we just recall informally that any event whose occurrence is not affected by changing the state of any finite collection of edges has probability either zero or one.

We call the existence of an infinite connected component event A. Note that A does not depend on the state of any finite collection of edges. Hence, it follows that $P_p(A)$ can only take the value zero or one. On the other hand we have $P_p(A) \geq \theta(p)$, so that $\theta(p) > 0$ implies $P_p(A) = 1$. Furthermore, by the union bound we have

$$P_p(A) \leq \sum_{x \in \mathbb{Z}^2} P_p(|C(x)| = \infty) = \sum_{x \in \mathbb{Z}^2} \theta(p), \tag{2.25}$$

so that $\theta(p) = 0$ implies $P_p(A) = 0$. $\qquad\square$

Before proceeding with the proof of Theorem 2.2.5, we give a sketch of the functions $\theta(p)$ and $\psi(p)$ in Figure 2.1. Note that the behaviour of $\theta(p)$ between p_c and one is not completely known, although it is believed to behave as a power law close to the critical point. Both functions are known to be continuous and their value at p_c is zero. It is also known that $p_c = 1/2$; this result was one of the holy grails in probability theory in the 1960s and 1970s and was finally proven by Kesten (1980), building upon a series of previous works of different authors. We give an outline of the proof in Chapter 4.

Proof of Theorem 2.2.5   This proof is based on a counting argument known as the *Peierls argument*, after Peierls (1936), who developed it in a different context, and it is divided into two parts. First we show that for $p < 1/3$, $\theta(p) = 0$. Then we show that for $p > 2/3$, $\theta(p) > 0$. The result then immediately follows by applying Theorem 2.2.2.

Let us start with the first part. We define a path as an alternating sequence of distinct vertices and edges that starts and ends with a vertex. The length of the path is the number of edges it contains. A circuit of length $n+1$ is a path of length n with one additional edge connecting the last vertex to the starting point. We first compute a bound on the total number of possible paths of length n departing from O in a fully connected lattice on $\mathbb{Z}^2$. This is a deterministic quantity, denoted by $\sigma(n)$, and satisfies

$$\sigma(n) \leq 4 \cdot 3^{n-1}, \tag{2.26}$$

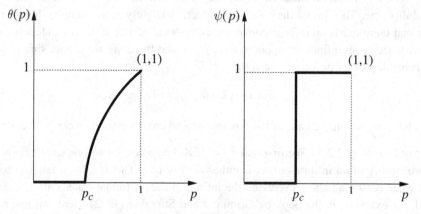

Fig. 2.1   Sketch of the phase transition.

because each step of a path on the lattice has at most three choices, apart from the first step that has four. We order the $\sigma(n)$ paths in some arbitrary way. Now let $N(n)$ be the number of paths of length n in our random grid, starting at O. Note that $N(n)$ is a random variable. If there exists an infinite path departing from the origin O, then for each n there must also exist at least one path of length n departing from O. Letting $I_i \in \{0, 1\}$ denote the indicator random variable of the existence of the ith path, in conjunction with the union bound this gives

$$\theta(p) \le P_p\left(N(n) \ge 1\right) = P_p\left(\bigcup_{i=1}^{\sigma(n)}\{I_i = 1\}\right)$$

$$\le \sum_{i=1}^{\sigma(n)} P_p(I_i = 1) = \sigma(n)p^n. \tag{2.27}$$

We now substitute the bound (2.26) into (2.27), obtaining

$$\theta(p) \le 4p \cdot (3p)^{n-1}, \text{ for all } n. \tag{2.28}$$

By choosing $p < 1/3$ and letting $n \to \infty$ the first part of the proof is complete.

The second part of the proof shows an application of the concept of a *dual lattice*. The idea is to evaluate the perimeter of a connected component using a dual lattice construction and then show that the probability that this is bounded is strictly less than one.

The dual lattice is defined by placing a vertex in each square of the lattice $\mathbb{Z}^2$, and joining two such vertices by an edge whenever the corresponding squares share a side; see Figure 2.2. We can also construct a dual of the random grid by drawing an edge in

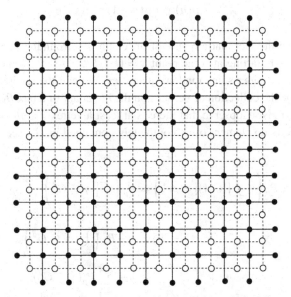

Fig. 2.2   A lattice and its dual, drawn with a dashed line.

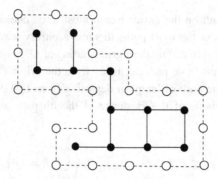

Fig. 2.3 The edges of a circuit in the dual surround any finite cluster in the original random grid.

the dual lattice, if it does not cross an edge of the original random grid, and deleting it otherwise. Note that any finite connected component in the random grid is surrounded by a circuit of edges in the dual random grid. Indeed, the edges of the dual block all possible paths to infinity of any finite cluster; see Figure 2.3. It follows that the statement $|C| < \infty$ is equivalent to saying that O lies inside a closed circuit of the dual. Kesten (1982) provides a surprisingly difficult rigorous proof of this statement that itself seems evident by inspecting Figure 2.3.

We start by looking at some deterministic quantities. Note that all the circuits in the dual lattice that contain the origin in their interior form a countable set $\mathcal{C}$ and let $\mathcal{C}_k \subset \mathcal{C}$ be the subset of them that surround a box of side k centred at O. Let the $\rho(n)$ be the number of circuits of length n of the dual lattice that surround the origin. This deterministic quantity satisfies

$$\rho(n) \leq n\sigma(n-1), \tag{2.29}$$

which follows from the fact that any circuit of length n surrounding the origin contains a path of length $n-1$ starting at some point $x = (k+1/2, 1/2)$ for some $0 \leq k < n$.

We now turn to consider some random quantities. We call the random grid G and its dual G^d. Let ∂B_k be the collection of vertices on the boundary of a box B_k of side length k centred at the origin. Now observe that there is at least a vertex $x \in \partial B_k$ with $|C(x)| = \infty$ if and only if there is no element of $\mathcal{C}_k$ completely contained in G^d, see Figure 2.4. This leads to

$$P_p\left(\bigcup_{x \in \partial B_k} \{|C(x)| = \infty\}\right) = P_p\left(\bigcap_{\gamma \in \mathcal{C}_k} \gamma \nsubseteq G^d\right)$$

$$= 1 - P_p\left(\bigcup_{\gamma \in \mathcal{C}_k} \gamma \subseteq G^d\right)$$

$$\geq 1 - \sum_{\gamma \in \mathcal{C}_k} P_p(\gamma \subseteq G^d), \tag{2.30}$$

using the union bound.

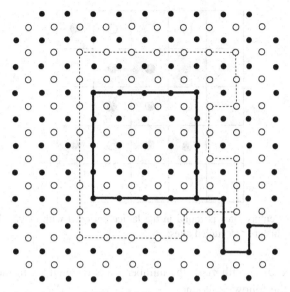

Fig. 2.4 There is at least a point on ∂B_k which lies on an infinite path if and only if there are no dual circuits surrounding B_k.

Furthermore, letting $q = 1 - p$, we note that a dual circuit of length n has probability of occurrence q^n, and exploiting the bounds (2.29) and (2.26) we have

$$\sum_{\gamma \in \mathcal{C}_k} P_p(\gamma \subseteq G^d) \leq \sum_{n=4k}^{\infty} n\sigma(n-1)q^n$$

$$\leq \frac{4}{9} \sum_{n=4k}^{\infty} (3q)^n n < 1, \tag{2.31}$$

where the last inequality holds by choosing $q < 1/3$, so that the series converges, and by choosing k large enough. Next, plug (2.31) into (2.30) and conclude that for k large enough,

$$P_p\left(\bigcup_{x \in \partial B_k} \{|C(x)| = \infty\} \right) > 0. \tag{2.32}$$

This clearly implies that for $q < 1/3$ we have $P(|C| = \infty) > 0$ and the proof is complete. $\qquad\square$

All the results we have derived up to now for the bond percolation model can also be derived for the site percolation model. The only change that is needed is in the application of the Peierls argument, in the second part of the proof of Theorem 2.2.5, which will lead to a different upper bound on p_c. The reason we need a slight modification is that there is no concept of dual lattice in site percolation. Accordingly, when we apply the Peierls argument we need to define a different lattice in which to look for circuits of empty sites that block paths to infinity of any occupied cluster. We do so by simply enriching the original lattice with diagonal connections between sites, see Figure 2.5. In this way we

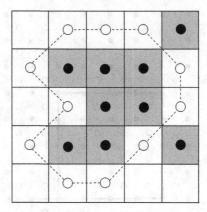

Fig. 2.5 A circuit of empty sites is found in the lattice enriched with diagonal connections that surrounds any finite cluster of occupied sites.

just need to replace $\sigma(n)$ with $\tau(n)$, the number paths of length n in the newly defined lattice, and to use the following bound

$$\tau(n) \leq 8 \cdot 7^{n-1}, \tag{2.33}$$

which holds since now each step of a path has at most seven choices, apart from the first step that has eight. We leave to the reader the computation of the actual upper and lower bounds on p_c that follows from this substitution inside the proof. We also mention that the exact value for p_c for site percolation on the square lattice is not known, however computer simulations suggest that it is close to $p_c \approx 0.59275$.

Bond and site percolation models can be easily extended to other network structures different from the grid. One can simply assign a probability to vertices or edges of any infinite connected graph. In order to obtain a non-trivial value of p_c, one then has to check if the Peierls argument can be applied. What is needed is an exponential bound on the probability of existence of a path of length n, and a suitable construction of circuits blocking finite components. It should be clear that not all infinite connected graphs lead to non-trivial values of p_c; see the exercises.

In general, we do not know the exact values of p_c on most graphs, but it turns out that p_c^{site} and p_c^{bond} satisfy a basic inequality on all graphs. The following theorem shows this relation and provides an example of a proof technique called *dynamic coupling* that will be often useful elsewhere. The idea is similar to the coupling we have already seen in the proof of Theorem 2.2.2, but in this case we give an algorithm that dynamically constructs realisations of two percolation processes along the way. These realisations are coupled, so that if two vertices are connected by a path in one realisation, they are also connected by a path in the other realisation. In this way, if the first process percolates, the other also does. The inequality in Theorem 2.2.7 is shown to be strict for a broad range of graphs by Grimmett and Stacey (1998).

Theorem 2.2.7 *For any infinite connected graph G we have that*

$$p_c^{site}(G) \geq p_c^{bond}(G). \tag{2.34}$$

Proof We want to describe a procedure that builds components of connected vertices in the site percolation process on G, and simultaneously components of connected vertices in the bond percolation process on G, in such a way that if there is an infinite cluster in the site model, then there must be an infinite cluster in the bond model as well. One aspect of the construction is that it is dynamic and creates a component along the way, beginning with a single vertex $x_0 \in G$.

We start by examining the edges that depart from x_0 and connect to nodes that we have not seen before. Each of these edges is marked dead, with probability $1 - p$, or it is marked as survivor otherwise. Each time an edge is marked, we also give the same mark to the node it connects to. Note that in this way each node spanning from x_0 survives independently with probability p, and at the same time this event is coupled with the outcome of the mark of its corresponding edge. In the next iteration, we move to one of the survivor nodes connected to x_0 and repeat the marking procedure to nodes that we have not seen before, in the same fashion. Then, we move to another survivor node, and so on. In this way, a tree spanning from x_0 is constructed. We make the following observations:

(i) Each time a new node is marked as survivor, it is in a connected component of survivor nodes and edges, rooted at x_0.
(ii) At each step of the procedure, conditioned on the state of all the nodes (edges) that have been marked in the past, each new node (edge) survives independently with probability p.

Let us now perform site percolation on the original graph G, by independently deleting nodes (and all edges emanating from them) from G, with probability $1 - p$, and focus on the resulting connected component centred at x_0. A way to create a realisation of this component is by using our marking algorithm on G. It follows that if $|C| = \infty$ in the site percolation model on G, then this is also true for the spanning tree obtained with our algorithm. But all nodes in this tree are also connected by survivor edges of G. Each edge in the spanning tree survives independently with probability p, and hence the tree is also a subgraph of the bond percolation component centred at x_0; it follows that this latter component is also infinite. We conclude that $p_c^{site} \geq p_c^{bond}$ and the proof is complete. □

2.3 Dependencies

An important extension of independent discrete percolation as considered so far, are models with dependencies between sites or edges. In this case, the state of each edge (site) can depend on the state of other edges (sites) of the graph.

We restrict ourselves here to *stationary* models, that is, models where the joint distribution of the state of any finite collection of edges (sites) does not change upon translations. In other words, the random graph has the same probabilistic behaviour everywhere; this should be compared with the notion of stationarity that we used in the construction of the Poisson process in Chapter 1.

The phase transition theorem generalises to these models, as long as edges (sites) that are separated by a path of minimum length $k < \infty$ on the original infinite graph, are independent. We give a proof in the case of discrete site percolation on the square lattice. This is easily extended to the bond percolation case and to other lattice structures different from the grid. In the following, distances are taken in L_1, the so-called *Manhattan distance*, that is the minimum number of adjacent sites that must be traversed to connect two points.

Theorem 2.3.1 *Consider an infinite square grid G, where sites can be either empty or occupied, and let $k < \infty$. Let p be the (marginal) probability that a given site is occupied. If the states of any two sites at distance $d > k$ of each other are independent, then there exist $p_1(k) > 0$ and $p_2(k) < 1$ such that $P(|C| = \infty) = 0$, for $p < p_1(k)$ and $P(|C| = \infty) > 0$, for $p > p_2(k)$.*

Note that in Theorem 2.3.1 there is no notion of a critical probability. Recall that in the independent case the existence of a unique critical value was ensured by the monotonicity of the percolation function, in conjunction with the upper and lower bounds that marked the two phases of the model. A dependent model might not even be characterised by a single parameter, and hence there is not necessarily a notion of monotonicity. However, we can still identify two different phases of the model that occur when the marginal probability of site occupation is sufficiently high, or sufficiently small. Note that the bounds we have given only depend on k and not on any further characteristics of the model.

Proof of Theorem 2.3.1 By looking at the proof of Theorem 2.2.5, we see all that is needed to apply the Peierls argument is an exponential bound on the probability of a path of occupied sites of length n. We refer to Figure 2.6. Consider a path starting at some site S_0, and let B_0 be the box of side length $2k + 1$ centred at S_0. The path must visit a site S_1 outside B_0 after at most $(2k + 1)^2$ steps. Note that the states of S_0 and S_1 are independent because their distance is greater than k. Now consider a square B_1 of the same size as B_0, centred at S_1. The path starting at S_0 visits a site S_2 outside $B_0 \cup B_1$ after at most $2(2k + 1)^2$ steps. Note that the states of S_0, S_1, S_2 are independent of each other. By iteration we have that in general the path starting at S_0 visits a new independent site S_i, outside $B(0) \cup B(1) \cup \cdots \cup B(i-1)$, after at most $i(2k + 1)^2$ steps. It follows that the total number of independent sites visited by a path of length n is at least $\lfloor n/(2k + 1)^2 \rfloor$. Hence the probability that such a path is completely occupied is at most

$$p^{\lfloor \frac{n}{(2k+1)^2} \rfloor}, \tag{2.35}$$

which gives the desired bound. $\qquad\square$

It is interesting to note that the bound obtained in this proof only depends on k; in the exercises the reader is asked to provide concrete numbers for $p_1(k)$ and $p_2(k)$. Note also that as $k \to \infty$ the bound tends to one, so the proof produces almost trivial estimates for p_1 and p_2 as dependencies tend to have longer range.

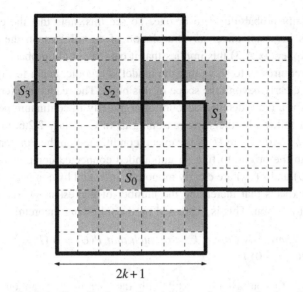

Fig. 2.6   Sites S_0, S_1, S_2, S_3 are independent of each other.

We conclude this section by pointing out that the percolation results we have seen so far can also be adapted to *directed* graphs. In this case, edges can be traversed only in one way. A typical model, for example, is directed site, or bond percolation on the square lattice, where all horizontal edges are oriented in one axis direction, while all vertical edges are oriented along the other axis direction. In this case one can define a critical probability $\vec{p}_c$ for directed percolation, i.e. for the existence of an infinite one way path departing from the origin. This critical value is larger than the critical probability of the undirected model p_c, but can still be bounded as $0 < \vec{p}_c < 1$. Details are given in, for instance, Kesten (1982).

2.4 Nearest neighbours; continuum percolation

We now turn to consider models in the continuum plane. One nice thing about these models is that we can use results from the discrete random networks to derive some of their percolation properties.

We start by considering the Poisson nearest neighbour model, where each point of a planar Poisson point process X is connected to its k nearest neighbours. As we have seen in Chapter 1, the density of the Poisson process does not play a role here, since by changing the unit of length we can vary the density of the process without changing the connections. In this sense, the model is *scale free* and we can assume the density to be as high as we want. Note also that in contrast to the previous models, in this case there is no independence between connections, as the existence of an edge in the network depends on the positions of all other points in the plane.

We call U the event that there exists an unbounded connected component in the resulting nearest neighbour random network. As in the discrete percolation model, this event can have only probability zero or one. To see this, note that the existence of an infinite cluster is invariant under any translation of coordinates on the plane, and by ergodicity (see Appendix A.3) this implies that it can have only probability zero or one. This, of course, requires as to show that the model is ergodic, and a formal treatment of ergodic theory is clearly beyond the scope of this book. The reader is referred to Meester and Roy (1996) for a more detailed account of ergodicity in continuum percolation.

In any case, it follows by the same reasoning as in Corollary 2.2.6, that $P(U) = 1$ is equivalent to $P(U_0) > 0$, where U_0 is the event, conditioned on the Poisson point process having a vertex in the origin, to find an unbounded connected component at the origin. Furthermore, $P(U)$ and $P(U_0)$ are clearly monotone in k, and by comparison with discrete percolation one expects that increasing the number of nearest neighbour connections k leads to a phase transition. This is expressed by the following theorem.

Theorem 2.4.1   *There exists a $2 \leq k_c < \infty$ such that $P(U) = 0$ $(P(U_0) = 0)$ for $k < k_c$, and $P(U) = 1$ $(P(U_0) > 0)$ for $k \geq k_c$.*

It is interesting to contrast this result with the one for the random tree given in Theorem 2.1. In that case the phase transition occurs when each node has at least one child on average, while in this case connecting to one nearest neighbour is not enough to obtain an infinite connected component. In this sense, we can say, quite informally, that trees are easier to percolate than nearest neighbour networks. We also mention that the exact value of k_c is not known, however Häggström and Meester (1996) have shown that $k_c = 2$ in high enough dimensions, and computer simulations suggest that $k_c = 3$ in two dimensions.

The proof of Theorem 2.4.1 is divided into two parts. The first part is of combinatorial flavour and exploits some interesting geometrical properties of 1-nearest neighbour models. Essentially, it rules out all the possible different geometric forms that an infinite connected component could possibly assume. The second part is a typical percolation argument based on *renormalisation* and coupling with a supercritical site percolation model on the square grid. Essentially, it shows that for k sufficiently large, discrete site percolation implies nearest neighbour percolation. Renormalisation arguments are of great value in statistical mechanics and their intuition of 'self similarity' of the space is very appealing. They are also sometimes applied non-rigorously to give good approximations of real phenomena. The basic idea is to partition the plane into blocks and look for *good* events occurring separately inside each block. The space is then renormalised by replacing each block with a single vertex. The state of each new vertex and the states of the edges connecting them are defined in terms of the good events occurring inside the corresponding blocks. In this way, the behaviour of the renormalised process can reduce to that of a known percolation process. One of the main difficulties when developing such a construction is that one has to be careful of not introducing unwanted dependencies in the renormalised process, while ensuring that neighbouring good renormalised boxes are somehow connected in the original process. The presence of such unwanted dependencies is what makes many of the heuristic reasonings not mathematically rigorous.

Proof of Theorem 2.4.1   For the first part of the proof it is sufficient to show that the model does not percolate for $k = 1$. Accordingly, we fix $k = 1$ and start by looking at the number of edges that can be incident to a Poisson point. This is called the *kissing number* of the nearest neighbour network and it is well known that there is a finite upper bound for it; see for example Zong (1998). It follows that if there exists an infinite connected component, there must also be an infinite path in the network and we want to rule out this possibility.

We introduce the following notation: call the nearest neighbour graph G and let G_d be an auxiliary graph where we represent connections using directed edges, writing $x \to y$ if y is the Poisson point nearest to x. This means that if there is an undirected edge between x and y in G, then there must either be $x \to y$, or $y \to x$, or both in G_d.

We have the following properties:

(i)   In a path of type $x \to y \to z$, with $x \neq z$, it must be that $|x - y| > |y - z|$, otherwise the nearest neighbour of y would be x instead of z, where $|\cdot|$ denotes Euclidean distance.

(ii) In G_d only loops formed by two edges are possible. That is because for any loop of type $x_1 \to x_2 \to \cdots \to x_n \to x_1$, $n > 2$, the following chain of inequalities must hold and produces a contradiction $|x_1 - x_2| > |x_2 - x_3| > \cdots > |x_n - x_1| > |x_1 - x_2|$.

(iii) Any connected component in G_d contains at most one loop, because otherwise for some point x of the component we would have $x \to y$ and $x \to z$, which is clearly impossible, because one node cannot have two nearest neighbours.

The situation arising from the three points (i)–(iii) above is depicted in Figure 2.7. It follows that we have to rule out only the two cases (1) an infinite component with one loop of length two, and (2) an infinite component without loops.

Let us look at the first possibility. We can assume that the Poisson process has density one. In order to reach a contradiction, let us assume that there are infinite components with one loop of length two. In this case, there is a certain positive probability that an arbitrary Poisson point is in a loop of length two, and is also contained in an infinite component. It follows that the expected number of such loops (that is, loops of length

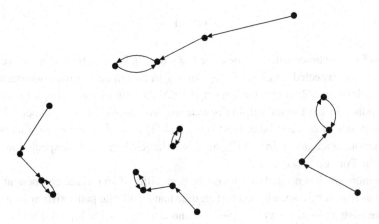

Fig. 2.7   Nearest neighbour clusters in G_d.

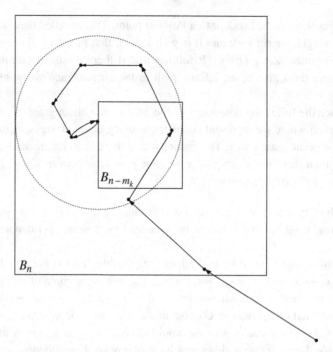

Fig. 2.8 The dashed disc has radius m_k and is centred at the node in the loop that has only one ingoing edge. We choose m_k so large that the probability that the k points nearest to the loop, in the infinite cluster, are inside the dashed disc is at least $1/2$.

two contained in an infinite component) in the box $B_n = [0, n] \times [0, n]$ is at least cn^2, for some positive constant c.

Next, for all $k > 0$ there is a number m_k such that the probability that the k points nearest to the loop in the infinite component of the loop, are all within distance m_k of the loop, is at least $1/2$; see Figure 2.8. Now choose k so large that

$$\frac{1}{4}kc > 1. \tag{2.36}$$

The reason for this choice will become clear in a moment. For this fixed k, we can choose n so large that the expected number of loops of length two in an infinite component inside B_{n-m_k} is at least $cn^2/2$. Since we expect more than half of these loops to have their nearest k points in the cluster within distance m_k, this implies that the expected number of Poisson points in B_n must be at least $(cn^2/2)(k/2) = kcn^2/4$. However, this is clearly a contradiction, since according to (2.36), this is larger than n^2, the expected number of points of the Poisson process in B_n.

A little more work is needed to rule out the possibility of an infinite component without loops. Let us reason by contradiction and assume that an infinite path exists at some $x_0 \in X$. That is, we have $x_0 \to x_1 \to x_2 \to x_3 \to \cdots$, and also $|x_0 - x_1| > |x_1 - x_2| > |x_2 - x_3| \cdots$. We proceed in steps to evaluate the probability of occurrence of such an infinite path.

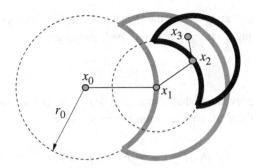

Fig. 2.9   The lightly highlighted line shows the boundary of the region where x_2 can fall to lie on the path after x_1. The dark highlighted line shows the boundary of the corresponding region for x_3.

Note that in the following procedure the Poisson process is constructed along the way, and we initially think of the whole space as being completely empty. We start with a point $x_0 \in X$ and grow a disc around x_0 until a point $x_1 \in X$ is found. This latter point is the nearest neighbour of x_0 and it is found with probability $P_1 = 1$. Now we need to find x_2 along the path. Note that the probability P_2 that x_2 is different from x_0 and x_1, corresponds to the existence of at least a point closer to x_1 than the distance between x_1 and x_0, but further from x_0 than x_1. Accordingly, writing $B(x, r)$ for the ball of radius r centred at x, we can grow a disc centred at x_1 until a point x_2 is found inside $A(B(x_1, r_0) \setminus B(x_0, r_0))$, where $A(\cdot)$ denotes area; see Figure 2.9. This latter point is found with probability

$$P_2 = 1 - e^{-A[B(x_1,r_0)\setminus B(x_0,r_0)]} \leq 1 - e^{-\pi r_0^2}. \tag{2.37}$$

By iterating the procedure in the natural way (see Figure 2.9 to visualise also the third step) we have the following recurrence at the generic step i: conditioned on the position of the first i points in the path,

$$P_{i+1} = 1 - e^{-A[B(x_i,r_{i-1})\setminus \cup_{j=0}^{i-1} B(x_j,r_j)]} \leq 1 - e^{-\pi r_0^2}. \tag{2.38}$$

Hence, given x_0, the probability of existence of the sequence $x_1, \ldots, x_i$ is bounded above by the product of the conditional probabilities

$$\prod_{j=1}^{i} P_j \leq (1 - e^{-\pi r_0^2})^{i-1} \tag{2.39}$$

which tends to zero as $i \to \infty$, since $r_0 > 0$. This immediately implies that $P(U_0) = 0$.

We can now prove the second part of the theorem. The objective here is to develop a renormalisation argument showing that discrete site percolation on the square grid implies nearest neighbour percolation, for sufficiently high values of k.

Let $0 < p_c < 1$ be the critical probability for site percolation on the square lattice. Let us consider a grid that partitions the plane into squares of side length one. Let us then further subdivide each of these unit squares into 7^2 subsquares of side length $1/7$ and let s_i denote one such subsquare. We denote by $X(s_i)$ the number of Poisson points falling inside s_i. We can assume that the density of the point process λ is so large that the probability of having no point inside one such subsquare is

$$P(X(s_i) = 0) < \frac{1 - p_c}{2 \cdot 7^2}. \tag{2.40}$$

We now consider the event A that there is at least one point inside each of the 7^2 subsquares of the unit square. By the union bound we have

$$P(A) \geq 1 - \sum_{i=1}^{7^2} P(X(s_i) = 0), \tag{2.41}$$

and by substituting inequality (2.40) into (2.41) we obtain

$$P(A) > 1 - 7^2 \frac{1 - p_c}{2 \cdot 7^2} = \frac{1 + p_c}{2}. \tag{2.42}$$

Next, choose an integer m so large that the probability of having more than $m/7^2$ points inside one subsquare is

$$P\left(X(s_i) > m/7^2\right) < \frac{1 - p_c}{2 \cdot 7^2} \tag{2.43}$$

and consider the event B that there are at most $m/7^2$ points inside each of the 7^2 subsquares of the unit square. By the union bound we have

$$P(B) \geq 1 - \sum_{i=1}^{7^2} P\left(X(s_i) > m/7^2\right) \tag{2.44}$$

and substituting the inequality (2.43) into (2.44) we obtain

$$P(B) > 1 - 7^2 \frac{1 - p_c}{2 \cdot 7^2} = \frac{1 + p_c}{2}. \tag{2.45}$$

From inequalities (2.42) and (2.45) we have that with probability greater than p_c each subsquare of the unit square contains at least one point *and* at most $m/7^2$ points, that is

$$P(A \cap B) > 1 - \left(1 - \frac{1 + p_c}{2}\right) - \left(1 - \frac{1 + p_c}{2}\right) = p_c. \tag{2.46}$$

The stage is now set for the coupling with the site percolation process. We call each unit square of the partitioned plane *good* if both events A and B occur inside it. Note that the event of a square being good is independent of all other squares and since $P(good) > p_c$ the good squares percolate. We want to show that this implies percolation in the Poisson nearest neighbour network of parameter m. To perform this last step, we focus on the subsquare placed at the centre of a good square; see Figure 2.10. Any point in an adjacent subsquare can be at most at distance $\sqrt{5}/7 < 3/7$ from any point inside the subsquare. Furthermore, no point inside such a subsquare has more than m points within distance $3/7$. This is because the entire good square contains at most m points. It follows that every point inside the subsquare at the centre connects to every point inside its adjacent subsquares, in an m-nearest neighbour model. By repeating the same reasoning we see that for any two adjacent good squares a path forms connecting all the points

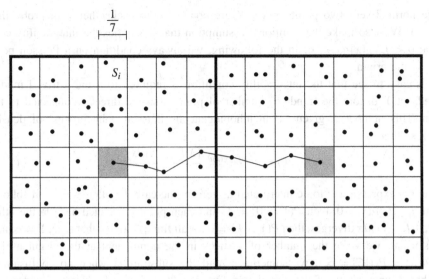

Fig. 2.10 Two adjacent *good* unit squares. Only one path connecting the subsquares at the centre is shown.

inside the subsquares at their centre. This shows that an unbounded component of good squares implies the existence of an unbounded connected component in the m-nearest neighbour model and the proof is complete.   □

We now make the following observation on the cluster site distribution that we shall use to prove other phase transition theorems: if the average cluster size is finite then the probability of it being infinite is zero. This may appear as a simple statement, but it is worth spending a few more words on it. As usual, we denote by $|C|$ the size of the cluster at the origin. We have

$$E(|C|) = \infty P(|C| = \infty) + \sum_{n=1}^{\infty} nP(|C| = n), \qquad (2.47)$$

where $0 \times \infty$ is defined as 0. From (2.47) it follows that

(i) $E(|C|) < \infty$ implies $P(|C| = \infty) = 0$;
(ii) $P(|C| = \infty) > 0$ implies $E(|C|) = \infty$;
(iii) $E(|C|) = \infty$ implies nothing.

It is therefore in principle possible, and worth keeping in mind, that the existence of the infinite cluster has probability zero, while the expected size of the cluster is infinite. Indeed this is for instance the case for some models at criticality.

2.5 Random connection model

We now consider the random connection model introduced in Chapter 1. Let X be a Poisson point process on the plane of density $\lambda > 0$. Let $g(\cdot)$ be a random connection function from $\mathbb{R}^2$ into $[0, 1]$ that depends only on the Euclidean norm $|x|$ and is non-increasing in the norm. Every two points $x, y \in X$ are connected to each other with probability $g(x - y)$. We also make the additional assumption that g satisfies the integrability condition: $0 < \int_{\mathbb{R}^2} g(x)dx < \infty$. In the following, we always condition on a Poisson point being at the origin.

It is easy to see that the integrability condition is required to avoid a trivial model. Indeed, let Y denote the (random) number of points that are directly connected to the origin. This number is given by an inhomogeneous Poisson point process of density $\lambda g(x)$, so that

$$P(Y = k) = e^{-\lambda \int_{\mathbb{R}^2} g(x)dx} \frac{[\lambda \int_{\mathbb{R}^2} g(x)dx]^k}{k!}, \tag{2.48}$$

where this expression is to be interpreted as zero in the case $\int_{\mathbb{R}^2} g(x)dx = \infty$. It follows that if $\int_{\mathbb{R}^2} g(x)dx = 0$, then $P(Y = 0) = 1$, and each point is isolated a.s. On the other hand, if $\int_{\mathbb{R}^2} g(x)dx$ diverges, then $P(Y = k) = 0$ for all finite k, and in that case, $Y = \infty$ a.s.

As usual, we write the number of vertices in the component at the origin as $|C|$ and $\theta(\lambda) = P_\lambda(|C| = \infty)$; we sometimes omit the subscript λ when no confusion is possible. Monotonicity of the percolation function $\theta(\lambda)$ should be clear: consider two random connection models with $\lambda_1 < \lambda_2$, thin the process of density λ_2 with probability $(1 - \lambda_1/\lambda_2)$, and follow the proof of Theorem 2.2.2 (see the exercises). We also have the following phase transition theorem.

Theorem 2.5.1 *There exists a $0 < \lambda_c < \infty$ such that $\theta(\lambda) = 0$ for $\lambda < \lambda_c$, and $\theta(\lambda) > 0$ for $\lambda > \lambda_c$.*

Two observations are in order. As usual, $P(|C| = \infty) > 0$ is equivalent to the existence a.s. of an unbounded connected component on $\mathbb{R}^2$. Furthermore, in virtue of the reasonings following (2.47), in the first part of the proof we will show that $\theta(\lambda) = 0$ by showing that $E(|C|) < \infty$, while in the second part of the theorem $\theta(\lambda) > 0$ implies $E(|C|) = \infty$.

Proof of Theorem 2.5.1 The first part of the proof constructs a coupling with the random tree model (branching process). One of the aspects of the construction is that the coupling is dynamic, and we create the point process along the way, so at the beginning of the construction, we think of the plane as being completely empty; compare with the proof of Theorem 2.4.1.

We start with a point x_0 in the origin, and consider the points directly connected to x_0. These form an inhomogeneous Poisson point process of density $\lambda g(x - x_0)$, which we call the first generation. We denote these (random) points by $x_1, x_2, \ldots, x_n$, ordered by modulus, say. In order to decide about the connections from x_1, we consider another inhomogeneous Poisson point process, independent from the previous one, and of density $\lambda g(x - x_1)(1 - g(x - x_0))$. The random points of this process represent points that are connected to x_1 but not connected to x_0. Similarly, the points of the second generation spanning from x_2 are obtained by an independent Poisson point process of density $\lambda g(x - x_2)(1 - g(x - x_1))(1 - g(x - x_0))$, representing points that are connected to x_2 but not connected to x_0 nor x_1. We now continue in the obvious way, at each point x_i spanning an independent Poisson point process and adding a factor $1 - g(x - x_j)$ with $j = i - 1$ to its density. When all the points of the second generation have been determined, we move to creating the next generation, this time excluding connections to all points already visited in all previous generations and generating new points along the way.

Note that the sequential construction described above produces a random graph G such that if any two points are connected to the origin in the random connection model, then they are also connected in G. However, the number of points at the nth generation of the construction is also bounded above by the number of points in the nth generation of a random tree of expected offspring $\mu = \lambda \int_{\mathbb{R}^2} g(x - x_n)dx = \lambda \int_{\mathbb{R}^2} g(x)dx$. That is because some connections in the construction are missing due to the additional factors $(1 - g(x - x_j)) < 1$. We can then choose λ small enough such that $\mu \leq 1$ and apply Theorem 2.1.1 to complete the first part of the proof.

For the second part of the proof we need to show that for λ large enough, $\theta(\lambda) > 0$. It is convenient, also in the sequel, to define $\bar{g} : \mathbb{R}^+ \to [0, 1]$ by

$$\bar{g}(|x|) = g(x), \tag{2.49}$$

for any $x \in \mathbb{R}^2$.

Let us partition the plane into boxes of side length 1. Note that the probability that any two Poisson points inside two adjacent boxes are connected by an edge, is at least $\bar{g}(\sqrt{5})$, since the diagonal of the rectangle formed by two adjacent boxes is $\sqrt{5}$. Furthermore, the probability that at least k points are inside a box can be made larger than $1 - \epsilon$, for arbitrarily small ϵ, by taking λ high. Let us focus on two adjacent boxes and let x_0 be a Poisson point in one of the two boxes. Hence for λ high enough, the probability that x_0 connects to at least one point in the adjacent box is given by

$$p \geq (1 - \epsilon) \left(1 - \left(1 - \bar{g}(\sqrt{5}) \right)^k \right). \tag{2.50}$$

Let us choose k and λ such that $p > p_c$, the critical probability for site percolation on the square lattice. We can now describe a dynamic procedure, similar to the one in Theorem 2.2.7, that ensures percolation in the random connection model. As usual, we construct a connected component along the way, starting with a point $x_0 \in X$. In the first iteration we determine the connections from x_0 to Poisson points in each of the four boxes adjacent to the one where x_0 is placed. We call each of these boxes occupied if there is at least a connection from x_0 to some point inside the box, empty otherwise. Note that these boxes are occupied independently, with probability $p > p_c$. In the second iteration we move to a point x_1 inside an occupied box directly connected to x_0, and examine the connections from x_1 to points in boxes that were never examined before. The procedure then continues in the natural way, each time determining the status of new boxes, independently with probability p, and spanning a tree rooted at x_0 along the way, that is a subgraph of the component centred at x_0 in the random connection model. Since $p > p_c$, the probability that the box of x_0 is in an unbounded connected component of adjacent boxes is positive, and this implies that x_0 is in an unbounded connected component of the random connection model with positive probability. □

We have seen that for every random connection function satisfying the integrability condition, there is a phase transition at some critical density value λ_c. It is natural to

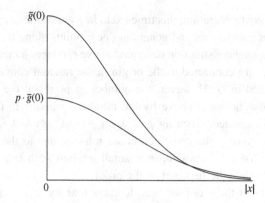

Fig. 2.11   Lowering and stretching the connection function $\bar{g}$.

ask how the value of λ_c changes with the shape of the connection function. A general tendency is that when the selection mechanism according to which nodes are connected to each other is sufficiently spread out, then a lower density of nodes, or equivalently on average fewer connections per node, are sufficient to obtain an unbounded connected component.

Let us consider a random connection function g (and associated $\bar{g}$) and some value $0 < p < 1$; define g_p by $g_p(x) = p \cdot g(\sqrt{p}x)$. This function, as illustrated in Figure 2.11, is a version of g in which probabilities are reduced by a factor of p, but the function is spatially stretched so as to maintain the same integral over the plane. Therefore, the expected number of connections of each Poisson point, $\lambda \int_{\mathbb{R}^2} g(x)dx$, is invariant under this transformation. We have the following result.

Theorem 2.5.2   *For a random connection model with connection function g, and $0 < p < 1$, we have*

$$\lambda_c(g) \geq \lambda_c(g_p). \tag{2.51}$$

Proof   The proof is based on an application of Theorem 2.2.7 and a coupling argument. We are to compare the critical densities associated with the connection functions g and g_p. We do this by relating both connection functions to a third connection function of larger effective area, namely $h_p(x) = g(\sqrt{p}x)$.

Consider a realisation $\mathcal{G}$ of a random connection model with density λ and connection function h_p. On $\mathcal{G}$, we can perform independent *bond* percolation with the same parameter p, by removing any connection (independent of its length) with probability $1 - p$, independently of all other connections. The resulting random graph can now effectively be viewed as a realisation of a random conection model with density λ and connection function $p.h_p(x) = g_p$.

On the other hand, we can also perform independent *site* percolation on $\mathcal{G}$ with connection function h_p, by removing any vertex of $\mathcal{G}$ (together with the connections emanating from it) with probability $1 - p$, independently of all other vertices. This results in a realisation of a random connection model with density $p\lambda$ and connection function

h_p, which can be seen (by scaling) as a realisation of a random connection model with density λ and connection function g; see the exercises.

We now apply Theorem 2.2.7 to $\mathcal{G}$: if site percolation occurs on $\mathcal{G}$, or equivalently, if a random connection model with density λ and connection function g percolates, then also bond percolation occurs, or equivalently, a random connection model with density λ and connection function g_p percolates. This proves the theorem. $\qquad\square$

Theorem 2.5.2 reveals a basic trade-off of any random connection model. The transformation considered essentially reduces the probability that two nodes form a bond. Edges, in some sense, are made prone to connection failures. The theorem shows that if at the same time we stretch the connection function so that the average number of connections per node remains the same, then such unreliable connections are at least as good at providing connectivity as reliable connections. Another way of looking at this is that the longer links introduced by stretching the connection function are making up for the increased unreliability of the connections.

Another way of spreading out the connection function is to consider only connections to nodes arbitrarily far away. Intuitively, if connections are spread out to the horizon, then it does not matter anymore where exactly nodes are located, as there is no notion of geometric distance scale. As the random connection model *loses its geometric component*, we also expect it to gain some independence structure and to behave similarly to an independent branching process. Since in branching processes the population may increase to infinity as soon as the expected offspring is larger than one, we expect the same to occur as nodes connect further away in the random connection model, at least approximately.

To visualise the spreading transformation we have in mind, consider shifting the function g outwards by a distance s, but squeeze the function after that, so that it maintains the same integral value over $\mathbb{R}^2$. Formally, for any $x \geq s$, we define $g_s(x) = g(c^{-1}(x-s))$, where the constant c is chosen so that the integral of g over the plane is invariant under the transformation; see Figure 2.12 for an illustrating example. Clearly, given $x_0 \in X$, as the shift parameter s is taken larger, the connections of x_0 are to Poisson points further

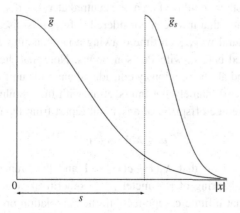

Fig. 2.12 Shifting and squeezing the connection function $\bar{g}$.

away and this, as discussed above, will bring geometric independence into the model, in the limit for $s \to \infty$. To ease the presentation of the results, we focus on shifting a rectangular connection function of unit area, that is

$$g(x) = \begin{cases} 1 & \text{if } |x| \le \sqrt{\frac{1}{\pi}}, \\ 0 & \text{if } |x| > \sqrt{\frac{1}{\pi}}. \end{cases} \tag{2.52}$$

The shifting operation can now be visualised as converting disc shapes into annuli shapes of larger radii over the plane. The general case of an arbitrary connection function is obtained following the same proof steps.

We denote by A_r the annulus with inner radius r and with area one, so that A_0 is just the disc of unit area. For each point x of the Poisson process, we consider the set $A_r(x) \equiv x + A_r$, that is, the annulus with inner radius r centred at x. We draw undirected edges between x and all points in $A_r(x)$. This gives a random network, and we denote its critical density value by $\lambda_c(r)$. Also note that since the area of each annulus is one, the density of the process also equals the expected number of connections per node. First, we show a strict lower bound on $\lambda_c(r)$.

Proposition 2.5.3 *For a random connection model with connection function g having value one inside an annulus of unit area and inner radius $r \ge 0$, and zero elsewhere, it is always the case that*

$$\lambda_c(r) > 1. \tag{2.53}$$

Proof We use a construction that is similar to the one in Theorem 2.5.1. Fix $r \ge 0$ and compare the percolation process to a branching process with a Poisson-λ offspring distribution as follows. We construct the percolation process by filling the plane with Poisson points incrementally, initially considering the plane as completely empty. We start by placing a point x_0 in the origin, we then take a Poisson-λ number of points, and we fill an annulus of area one centred in the origin, by distributing them uniformly and independently inside this annulus. Points in the annulus are then directly connected to x_0. Subsequent children of each point x inside the annulus are also distributed uniformly and independently over another annulus of area one centred at x, but if such a child happens to fall into one of the annuli that has been considered before, it is discarded. The procedure then iterates in the natural way, each time drawing a new annulus and filling some part of it, that was not filled before, with Poisson points. Note that the overlap between an annulus centred at x and all the previously considered annuli is uniformly bounded below by some number $c(r) > 0$, namely the intersection with the annulus of the parent of x. This means that the average offspring of any point (apart from the origin) is

$$\mu \le \lambda(1 - c(r)). \tag{2.54}$$

Hence, there is a $\lambda_0 > 1$ so that $\lambda_0(1 - c(r)) < 1$ and the branching process centred at x_0 and of Poisson offspring of parameter λ_0 is subcritical, and hence dies out. This immediately implies that infinite components in the percolation process cannot exist for λ_0, which shows that $\lambda_c(r)$ is strictly larger than 1. □

We have shown that the number of connections per node needed for percolation in the annuli random connection model is always greater than one, the critical offspring of a branching process. We now show that this lower bound is achieved asymptotically as the radius of the annuli tends to infinity, and connections are spread out arbitrarily far over the plane. This means that eventually, as connections are more spread out in the random connection model, on average one connection per node is enough to percolate.

Theorem 2.5.4 *For a random connection model with connection function g having value one inside an annulus of unit area and inner radius $r \geq 0$, and zero elsewhere, we have that*

$$\lim_{r \to \infty} \lambda_c(r) = 1. \tag{2.55}$$

The proof of this theorem is quite involved and it uses many of the results that we have seen so far. Therefore, it is a good exercise to review some of the techniques we have learned. First, there is a coupling with the random tree model (branching process), but this coupling is used only for a bounded number of generations. Then, the process is renormalised and coupled with directed site percolation on the square lattice, which eventually leads to the final result.

Proof of Theorem 2.5.4 We first describe a supercritical spatial branching process which is, in some sense to be made precise below, the limiting object of our percolation process as $r \to \infty$.

A spatial branching process. Consider an ordinary branching process with Poisson-λ offspring distribution, where $\lambda > 1$. This process is supercritical, and hence there is a positive probability that the process does not die out. We add a geometric element to this process as follows: The 0th generation point is placed at the origin, say. The children of any point x of the process are distributed uniformly and independently over the circumference of a circle with radius one, centred at x.

A sequential construction of the percolation process. We now describe a way to construct a percolation cluster in our percolation process, which looks very much like the branching process just described. We will then couple the two processes. One of the aspects of this construction is that we create the point process along the way, so at the beginning of the construction, we think of the plane as being completely empty. The density of the underlying Poisson process is the same $\lambda > 1$ as above.

We start with a point in the origin, and consider the annulus $A_r = A_r(0)$. We now 'fill' A_r with a Poisson process, that is, we take a Poisson-λ random number of points, and distribute these uniformly (and independently of each other) over A_r. These points are directly connected to the origin. If there are no points in A_r we stop the process; if there are points in A_r we denote these by $y_1, y_2, \ldots, y_s$, ordered by modulus, say. In order to decide about the connections from y_1, we consider $A_r(y_1)$ and 'fill' this annulus with an independent Poisson process, in the same way as before. The (random) points that we obtain in $A_r(y_1)$ are directly connected to y_1 but not to 0. Now note that we make a mistake by doing this, in the sense that the region $A_r(0) \cap A_r(y_1)$ is not empty, and this region has now been filled twice, and therefore the intensity of the Poisson process in the intersection is 2λ instead of the desired λ. For the moment we ignore this problem; we come back to this in a few moments. We now continue in the obvious way, each time 'filling' the next annulus with a Poisson process, and each time possibly making a mistake as just observed.

Coupling between branching process and percolation process. Ignoring the mistakes we make, the sequential construction described above is similar to the spatial branching process. We can actually couple the two processes (still ignoring mistakes) by insisting that the offspring of the branching process also be the points of the percolation process. If a point in the branching process is placed at a certain position (at distance one) from its parent, then the point in the percolation process is located at the same relative angle, and uniformly distributed over the width of the annulus. Since $\lambda > 1$, the percolation process would continue forever with positive probability, thereby creating an infinite percolation component.

However, we have to deal with the mistakes we make along the way. We have two tools at our disposal that can be helpful now. First, it should be noted that the overlap between the various annuli gets smaller as $r \to \infty$. Second, we will only use the coupling between the spatial branching process and the percolation process for a uniformly bounded number of annuli, and then build a renormalised process which we couple with a supercritical directed site percolation on a square lattice.

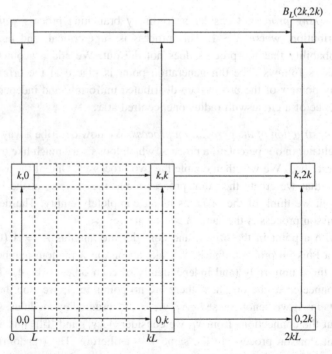

Fig. 2.13 Renormalisation blocks and coupling with directed discrete site percolation. We divide the positive quadrant into boxes of size $L \times L$. A directed site percolation model is constructed selecting boxes that are at $(k-1)L$ distance apart.

Renormalisation blocks. We now describe the coupling (limited to a uniformly bounded number of annuli) and the renormalisation. We first look at a construction for the spatial branching process, and then show that the same construction is achieved in the percolation process, with arbitrarily large probability. We refer to Figure 2.13. Divide the positive quadrant into boxes of size $L \times L$, where we choose L in a moment. The box with lower left-most point (iL, jL) is denoted by $B_L(i, j)$. Let ϵ and δ be given positive numbers, and let λ be as before.

We consider N spatial branching processes that evolve in parallel, starting from box $B_L(0, 0)$, and place a total ordering on the progeny we observe such that $x_{ab} < y_{cd}$ if $(a < c)$, or $(a = c$ and $b < d)$, where a, c represent the generation numbers of children x and y respectively, and b, d represent their Euclidean distances from an arbitrarily chosen origin. We now choose various quantities as follows.

(i) First choose N so large that the probability that at least one out of a collection of N independent spatial branching processes survives forever, is at least $1 - \epsilon$.

(ii) Then choose L so large that the probability that the box $B_L(0, 0)$ contains a collection of N points of the Poisson process of intensity λ such that no two points of the collection are within distance δ of each other, is at least $1 - \epsilon$. We call such a collection of N points a *good* collection.

(iii) Then choose k and M so large that in the spatial branching process (which, we recall, uses circles of radius one) the following is the case: if we start with any good collection of points in $B_L(0, 0)$, and we discard all further offspring of any point which falls in either $B_L(k, 0)$ or $B_L(0, k)$, then the probability that the total progeny of this collection, restricted to the first M points, contains a good collection in both $B_L(k, 0)$ and $B_L(0, k)$, is at least $1 - 4\epsilon$. The possibility of this choice requires a little reflection. We want to ensure that the N branching processes, after generating at most M points, will create a good collection of points in the two 'target' boxes $B_L(k, 0)$ and $B_L(0, k)$, even if we discard all offspring departing from points inside the two target boxes. Among the initial N branching processes starting in $B_L(0, 0)$, there is at least one that survives forever with high probability. By taking the distance factor k large enough we can also ensure with high probability that this surviving process generates an arbitrarily large collection of points before ever reaching any of the two target boxes. Each of these intermediate points has positive probability of having an infinite line of descendants. Since a single line of descent of any point follows a simple two-dimensional random walk with zero drift, this random walk is recurrent, and it will end up in either $B_L(0, k)$ or $B_L(k, 0)$. The probability that this happens for at least N lines of descent in each of the two target boxes and that the collection of 'terminal' points in each of the two target boxes contain a good set, can be made arbitrarily high provided that the number of intermediate starting points is high enough. Finally, the probability that this happens in a uniformly bounded number of generated points can be as high as we like by taking the allowed total number M of points large enough.

(iv)   Finally, we choose a δ' small enough so that the probability that the distance
       between any two of the first M points generated by the initial N branching processes
       is smaller than δ', is at most ϵ.

Note that the construction described up to now has been in terms of the spatial branching
process and it ensures that a good collection of points in $B_L(0,0)$ can create good collec-
tions in both $B_L(0,k)$ and $B_L(k,0)$, in a bounded number of iterations, with probability
at least $1 - 4\epsilon$. We now want to show that it is also possible to obtain the same, with
high probability, in the sequential percolation process. To do this we will need to take
the radius r of the annuli in the percolation process large enough. First of all, we note
that if we fix an upper bound M of annuli involved, and $\epsilon > 0$, we can choose r so large
that the probability that in the union of N sequential percolation processes, any point falls
into an intersection of two among the first M annuli, is at most ϵ. This is because we
start the N processes with annuli separated by at least δ, and evolve, generating a total
number of at most M annuli that are at distance at least δ' to each other. Hence, the total
overlap between the annuli can be made as small as we want by taking r large.

   The percolation process and the branching process now look alike in the first M
steps, in the sense that if the branching process survives while generating M points, the
percolation process also survives with high probability. To complete the construction we
need something slightly stronger than this. We also need to make sure that if a point
in the branching process ends up in a certain box $B_L(i,j)$, then the corresponding point
in the percolation process ends up in the corresponding box $B_{rL}(i,j)$ (the box with side
length rL whose lower left corner is at (irL, jrL)), and vice versa. Note that since the
annuli have a certain width, two offspring of the same parent will not be at the exact
same distance from the parent. Therefore, points can possibly end up in the wrong box.
However, the probability that there is a point which ends up in the wrong box can again be
made less than ϵ by taking r large. To explain why this is, note that the spatial branching
process has distance one between a parent and child, and the choice of N, L, M and δ'
are in terms of this process. When we couple the branching process with the percolation
process and we take r large, we also have to scale the whole picture by a factor r. When
we do this, the width of each annulus becomes smaller and tends to zero. Therefore, the
probability of making a mistake by placing a point in the wrong box decreases to zero
as well.

Dynamic coupling with discrete percolation. We are now ready to show that the renormalisation described above can be coupled with a supercritical directed site percolation
process on a square lattice. Let us order the vertices corresponding to boxes in the positive
quadrant in such a way that the modulus is non-decreasing. We look at vertices (i,j). We
call the vertex $(0,0)$ *open* if the following two things happen in the percolation sequential
construction:

(i)   The box $B_{rL}(0,0)$ contains a good collection of points; we choose one such collection
       according to some previously determined rule.
(ii)   The progeny of this chosen good collection, restricted to the first M annuli of the
       process (and where we discard further offspring of points in any of the two target

boxes $B_{rL}(0,k)$ and $B_{rL}(k,0)$) contains a good collection in both $B_{rL}(0,k)$ and $B_{rL}(k,0)$.

We now consider the vertices (i, j) associated with boxes of the first quadrant separated by distance kL one by one, in the given order. The probability that $(0, 0)$ is open can be made as close to one as desired, by appropriate choice of the parameters. In particular, we can make this probability larger than $\vec{p}_c$, where $\vec{p}_c$ is the critical value of directed two-dimensional independent site percolation on the square lattice.

If the origin is not open, we terminate the process. If it is open, we consider the next vertex, $(0, k)$ say. The corresponding box $B_{rL}(0, k)$ contains a good collection, and we can choose any such good collection according to some previously determined rule. We start all over again with this good collection of points, and see whether or not we can reach $B_{rL}(k, k)$ and $B_{rL}(0, 2k)$ in the same way as before. If this is the case, we declare $(0, k)$ open, otherwise we call it closed. Note that there is one last problem now, since we have to deal with overlap with annuli from previous steps of the algorithm, that is, with annuli involved in the step from $(0, 0)$ to $(0, k)$. This is easy though: since we have bounded the number of annuli involved in each step of the procedure, there is a uniform upper bound on the number of annuli that have any effect on any given step of the algorithm. Therefore, the probability of a mistake due to any of the previous annuli can be made arbitrarily small by taking r even larger, if necessary. This shows that we can make the probability of success each time larger than $\vec{p}_c$, no matter what the history of the process is. This implies that the current renormalised percolation process is supercritical. Finally, it is easy to see that if the renormalised process percolates, so does the underlying percolation process, proving the result. $\qquad\square$

It is not hard to see that this proof can be generalised to different connection functions g. In the general case, the offspring of a point is distributed according to an inhomogeneous Poisson process, depending on the connection function. Hence the following theorem.

Theorem 2.5.5 *For a random connection model of connection function g, such that $\int_{\mathbb{R}^2} g(x)dx = 1$, we have*

$$\lim_{s \to \infty} \lambda_c(g_s) = 1. \tag{2.56}$$

We end our treatment on phase transition in random connection models by looking at one effect that is somehow the opposite of spreading out connections: we consider the random connection model in the high density regime. This means that we expect a high number of connections between nodes that are closely packed near each other. Of course, as $\lambda \to \infty$, by (2.48) each Poisson point tends to have an infinite number of connections, and hence $\theta(\lambda) \to 1$. It is possible, however, to make a stronger statement regarding the rate at which finite components, of given size k, disappear.

Theorem 2.5.6 *In a random connection model at high density, points tend to be either isolated, or part of an infinite connected component. More precisely,*

$$\lim_{\lambda \to \infty} \frac{-\log(1 - \theta(\lambda))}{\lambda \int_{\mathbb{R}^2} g(x)dx} = 1. \tag{2.57}$$

Note that the theorem asserts that $1 - \theta(\lambda)$ behaves as $\exp(-\lambda \int_{\mathbb{R}^2} g(x)dx)$, which is the probability of a Poisson point being isolated. In other words, the rate at which $\theta(\lambda)$ tends to one corresponds exactly to the rate at which the probability of being isolated tends to zero. It follows that finite components of size $k > 1$ tend to zero at a higher rate, and at high densities all we see are isolated points, or points in the infinite cluster. We do not give a full proof of Theorem 2.5.6 here, but we give some intuition on why finite components tend to be isolated nodes at high densities, describing the phenomenon of *compression* of Poisson points in the next section.

2.6 Boolean model

All the results given for the random connection model also hold in the special case of the boolean model. It is interesting, however, to restate the phase transition theorem, and emphasise the scaling properties of the Poisson process. We define the node degree of the random boolean network as the average number of connections of a point of the Poisson process, given by $\xi = 4\pi r^2 \lambda$, and we give the following three equivalent formulations of the phase transition.

Theorem 2.6.1

(i) *In a boolean random network of radius r, there exists a critical density $0 < \lambda_c < \infty$ such that $\theta(\lambda) = 0$ for $\lambda < \lambda_c$, and $\theta(\lambda) > 0$ for $\lambda > \lambda_c$.*

(ii) *In a boolean random network of density λ, there exists a critical radius $0 < r_c < \infty$ such that $\theta(r) = 0$ for $r < r_c$, and $\theta(r) > 0$ for $r > r_c$.*

(iii) *In a boolean random network, there exists a critical node degree $0 < \xi_c < \infty$ such that $\theta(\xi) = 0$ for $\xi < \xi_c$, and $\theta(\xi) > 0$ for $\xi > \xi_c$.*

Although exact values of the critical quantities in Theorem 2.6.1 are not known, analytic bounds can be easily obtained adapting the proof of Theorem 2.5.1, and computer simulations suggest that $\xi_c = 4\pi r_c^2 \lambda_c \approx 4.512$.

The proof of Theorem 2.6.1 follows immediately from Theorem 2.5.1 and the following proposition.

Proposition 2.6.2 *In a boolean random network it is the case that*

$$\lambda_c(r) = \frac{\lambda_c(1)}{r^2}. \tag{2.58}$$

Proof Consider a realisation G of the boolean random network with $r = 1$. Scale all distances in this realisation by a factor r, obtaining a scaled network G_s. One can see G_s as a realisation of a boolean model where all discs have radius $1/r$, and the density of the Poisson process is $\lambda(1)/r^2$. However, the connections of G and G_s are the same, and this means that if G percolates, G_s also does so. On the other hand, if G does not percolate, G_s does not either. It follows that $\lambda_c(G_s) = \lambda_c(1)/r^2$, which concludes the proof. $\square$

We now turn to the compression phenomenon. We have seen in Theorem 2.5.6 that in any random connection model at high density, finite clusters tend to be formed by isolated points. These are the last finite components to eventually disappear, as $\lambda \to \infty$. In the special case of a boolean model we can give some additional geometric intuition on what happens.

For large λ, $P(|C| = k)$ is clearly very small for any fixed $k \geq 0$. Note that a necessary and sufficient condition for this event to occur is to have a component of k connected discs surrounded by a fence of empty region not covered by discs. By referring to Figure 2.14, we see that this corresponds to having the region inside the highlighted dashed line not containing any Poisson point other than the given k, forming the isolated cluster of discs. Clearly, the area of this region is smaller when the k points are close together. Hence, in a high density boolean model $\{|C| = k\}$ is a rare event, but *if* it occurs, it is more likely in a configuration where the k points collapse into a very small region, and the approximately circular-shaped area around them of radius $2r$ is free of Poisson points.

To make more precise considerations on the compression phenomenon, consider the following sufficient condition to have an isolated component of $k + 1 < \infty$ points: condition on a point being at the origin, and let, for $\alpha < r$, $S = S_\alpha$ be the event that a disc of radius α contains k additional points, and an annulus outside it of width $2r$ does not contain any point of the Poisson process; see Figure 2.15. Note that if S occurs, then there is an isolated cluster of size $k + 1$ at the origin, because the boundary of the empty region

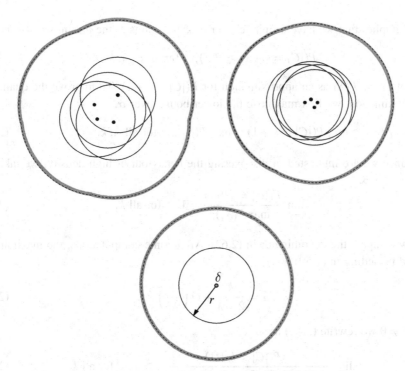

Fig. 2.14 The compression phenomenon.

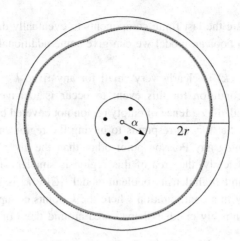

Fig. 2.15   Sufficient condition for an isolated component.

required around the k points for them to be isolated is contained inside the annulus. The probability of the event S can be computed as follows:

$$P(S) = \frac{(\lambda \pi \alpha^2)^k}{k!} e^{-\lambda \pi \alpha^2} e^{-\lambda(\pi(\alpha+2r)^2 - \pi \alpha^2)}$$

$$= \frac{(\lambda \pi \alpha^2)^k}{k!} e^{-\lambda \pi (\alpha+2r)^2}. \tag{2.59}$$

Since S implies that we have a finite cluster of $k+1$ points at the origin, we also have

$$P(|C| = k+1) \geq P(S), \quad \text{for all } \alpha, \lambda, k. \tag{2.60}$$

We want to use $P(S)$ as an approximation for $P(|C| = k+1)$. To improve the quality of the approximation, we first maximise the lower bound over α,

$$P(|C| = k+1) \geq \max_\alpha P(S_\alpha), \quad \text{for all } \lambda, k. \tag{2.61}$$

Then, since we are interested in discovering the behaviour at high density, we take the limit for $\lambda \to \infty$,

$$\lim_{\lambda \to \infty} \frac{P(|C| = k+1)}{\max_\alpha P(S_\alpha)} \geq 1, \quad \text{for all } k. \tag{2.62}$$

We now compute the denominator in (2.62). After some computations, the maximum is obtained by setting in (2.59)

$$\alpha = \frac{k}{2\pi r \lambda} + O\left(\frac{1}{\lambda^2}\right). \tag{2.63}$$

When $k \neq 0$ we rewrite (2.62) as

$$\lim_{\lambda \to \infty} \frac{P(|C| = k+1)}{\exp\left(-\lambda \pi (2r)^2 - k \log \frac{\lambda}{k} - O(1)\right)} \geq 1, \quad \text{for all } k. \tag{2.64}$$

Now, note that the bound is tight for $k = 0$. In this case the denominator of (2.62) behaves as $e^{-\lambda \pi (2r)^2}$, which is exactly the probability that a node is isolated, and which appears at the numerator. It turns out that the bound is indeed tight for all values of k. To see why is this so, let us look again at Figure 2.15 representing the sufficient condition S. In the limit for $\lambda \to \infty$, by (2.63) α tends to zero, and the annulus in the figure becomes a disc of radius $2r$. But remember that the disc of radius α has to contain all $k + 1$ points of the finite component centred at O. This means that the k additional Poisson points inside it must be arbitrarily close to each other as $\alpha \to 0$. Recall now what happens to the *necessary* condition for k points to be isolated, when these points become close to each other: the empty region *required* for the finite component to be isolated becomes exactly a circle of radius $2r$, see Figure 2.14. It follows that when $\alpha \to 0$, the sufficient condition to have $\{|C| = k + 1\}$ becomes geometrically the same as the necessary condition, and our corresponding approximation (2.64) becomes tight. This latter reasoning can be made more rigorous with some additional work; see Meester and Roy (1996).

Finally, by looking at the denominator of (2.64), we see that the probability of finite components tends to zero at a higher rate for larger values of k. This explains why isolated nodes are the last ones to disappear as $\lambda \to \infty$ and we have the equivalent of Theorem 2.5.6 for the boolean model.

Theorem 2.6.3

$$\lim_{\lambda \to \infty} \frac{1 - \theta(\lambda)}{e^{-\lambda \pi (2r)^2}} = 1. \tag{2.65}$$

2.7 Interference limited networks

We now want to treat the interference network model. As we shall see, there are some additional technical difficulties to show the phase transition in this model, due to its *infinite range* dependence structure. Furthermore, the model presents a number of interesting properties besides the phase transition. We start this section by stating two technical theorems that will turn out to be useful in the proofs.

The first theorem shows a property of the boolean model, namely that in the super-critical regime there are with high probability paths that cross a large box in the plane.

Theorem 2.7.1 *Consider a supercritical boolean model of radius r and density $\lambda > \lambda_c$. For any $0 < \delta < 1$, let $R_{\delta n}$ be a rectangle of sides $\sqrt{n} \times \delta \sqrt{n}$ on the plane. Let $R_{\delta n}^{\leftrightarrow}$ denote the event of a left to right crossing inside the rectangle, that is, the existence of a connected component of Poisson points of $R_{\delta n}$, such that each of the two smaller sides of $R_{\delta n}$ has at least a point of the component within distance r from it. We have*

$$\lim_{n \to \infty} P(R_{\delta n}^{\leftrightarrow}) = 1. \tag{2.66}$$

A proof of this theorem can be found in the book by Meester and Roy (1996) (Corollary 4.1); see also Penrose (2003). We shall prove a similar property in the context of discrete percolation in Chapter 5.

The second result we mention is known as *Campbell's theorem*. We prove a special case here.

Theorem 2.7.2 **(Campbell's theorem)** *Let X be a Poisson process with density λ, and let $f : \mathbb{R}^2 \to \mathbb{R}$ be a function satisfying*

$$\int_{\mathbb{R}^2} \min(|f(x)|, 1)\, dx < \infty. \tag{2.67}$$

Define

$$\Sigma = \sum_{x \in X} f(x). \tag{2.68}$$

Then we have

$$E(\Sigma) = \lambda \int_{\mathbb{R}^2} f(x)dx \tag{2.69}$$

and

$$E(e^{s\Sigma}) = \exp\left(\lambda \int_{\mathbb{R}^2} (e^{sf(x)} - 1)dx \right), \tag{2.70}$$

for any $s > 0$ for which the integral on the right converges.

Proof To see why the first claim is true, consider a special function f, namely $f(x) = \mathbf{1}_A(x)$, the indicator function of the set A. Then Σ is just the number of points in A, and the expectation of this is $\lambda|A|$ which is indeed equal to $\lambda \int_{\mathbb{R}^2} f(x)dx$. For more general functions f one can use standard approximation techniques from measure theory.

To prove the second claim, consider a function f that takes only finitely many non-zero values $f_1, f_2, \ldots, f_k$ and which is equal to zero outside some bounded region. Let, for $j = 1, \ldots, k$, A_j be defined as

$$A_j = \{x; f(x) = f_j\}. \tag{2.71}$$

Since the A_j are disjoint, the random variables $X_j = X(A_j)$ are independent with Poisson distributions with respective parameters $\lambda|A_j|$. Furthermore, we have that

$$\Sigma = \sum_{j=1}^k f_j X_j. \tag{2.72}$$

In general, for any Poisson random variable X with parameter μ and $s \in \mathbb{R}$ we can write

$$E(e^{sX}) = \sum_{k=0}^{\infty} e^{-\mu} \frac{\mu^k}{k!} e^{sk}$$

$$= e^{-\mu} \sum_{k=0}^{\infty} \frac{(\mu e^s)^k}{k!}$$

$$= e^{\mu(e^s - 1)}. \tag{2.73}$$

Using this, we may now write

$$E(e^{s\Sigma}) = \prod_{j=1}^{k} E(e^{sf_j X_j})$$

$$= \prod_{j=1}^{k} \exp\left(\lambda |A_j|(e^{sf_j} - 1)\right)$$

$$= \exp\left(\sum_{j=1}^{k} \int_{A_j} \lambda(e^{sf(x)} - 1)dx\right)$$

$$= \exp\left(\int_{\mathbb{R}^2} \lambda(e^{sf(x)} - 1)dx\right). \tag{2.74}$$

This proves the result for this special class of functions f. In order to prove it for general f, one uses some standard approximation techniques from measure theory; see for instance Kingman (1992), pages 29–30. □

We are now ready to discuss some properties of interference limited networks. We construct a random network as follows. Let X be a Poisson point process on the plane of density $\lambda > 0$. Let $\ell : \mathbb{R}^2 \times \mathbb{R}^2 \to \mathbb{R}$, be such that $\ell(x, y) = \ell(y, x)$, for all $x, y \in \mathbb{R}^2$; and let P, T, N, be positive parameters and γ non-negative. For each pair of points $x_i, x_j \in X$, define the ratio

$$SNIR(x_i \to x_j) = \frac{P\ell(x_i, x_j)}{N + \gamma \sum_{k \neq i,j} P\ell(x_k, x_j)}, \tag{2.75}$$

and place an undirected edge between x_i and x_j if both $SNIR(x_i \to x_j)$ and $SNIR(x_j \to x_i)$ exceed the threshold T. As usual, we say that the model is supercritical, i.e., it percolates if $P(|C| = \infty) > 0$, where $|C|$ indicates the number of points in the cluster at the origin.

Note that if $\gamma = 0$ then (2.75) reduces to (1.14) and the model behaves as a standard boolean model which percolates for $\lambda > \lambda_c$. We will show that percolation occurs for all $\lambda > \lambda_c$, by taking $\gamma(\lambda) > 0$ sufficiently small. We will also show that for any fixed γ, increasing the density λ of the Poisson point process always leads to a subcritical regime.

Before proceeding further and formally describing these results, it is of help to visualise them by looking at Figure 2.16. Numerical simulations show the value of γ^* below which the network percolates. In practice, γ^* marks the boundary of a supercritical region, that can be entered for given values of λ and γ. Note that γ^* becomes positive for $\lambda > \lambda_c$, and it tends to zero as $\lambda \to \infty$. Moreover, γ^* appears to be uniformly bounded from above. Next, we put these observations into a rigorous framework.

We start by making the following natural assumptions on the attenuation function $\ell(\cdot)$:

(i) $\ell(x, y)$ only depends on $|x - y|$, that is, $\ell(x, y) = l(|x - y|)$ for some function $l : \mathbb{R}^+ \to \mathbb{R}^+$;
(ii) $\int_y^\infty x l(x) dx < \infty$ for some $y > 0$;
(iii) $l(0) > TN/P$;
(iv) $l(x) \leq 1$, for all $x \in \mathbb{R}$;
(v) l is continuous and strictly decreasing on the set where it is non-zero.

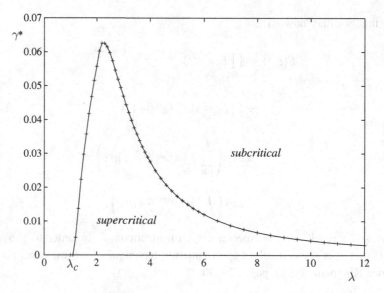

Fig. 2.16 The curve shows the critical value of γ^* below which the network percolates. The parameters of this simulation are $T = 1$, $N = 10^4$, $P = 10^5$, $l(x) = \max(1, x^{-3})$.

The first assumption is very natural, stating that the attenuation function depends only on the Euclidian distance between two points. The second and third assumptions are needed for the existence of links: we need (ii) to bound the interference and ensure convergence of the series in (2.75); on the other hand, we need (iii) to ensure that enough power is received to establish communication. The last two assumptions have been introduced for mathematical convenience, but also make sense in the physical world. The above assumptions immediately imply that the length of the edges in the resulting random network is uniformly bounded by $l^{-1}(TN/P)$. This of course implies that each node has a finite number of neighbours, since all of its connections have bounded range and $\lambda < \infty$, but it does not imply that this number can be uniformly bounded. For example, in the case of a boolean network ($\gamma = 0$), the number of neighbours of a given node cannot be uniformly bounded, however they are all at a bounded distance $2r$ from it. The next proposition shows that when $\gamma > 0$, indeed a uniform bound on the number of neighbours holds. Moreover, it also gives an upper bound $\gamma < 1/T$ required for the model to percolate.

Proposition 2.7.3 *For $\gamma > 0$ any node $x \in X$ is connected to at most $1 + \frac{1}{\gamma T}$ neighbours.*

Proof For all $x \in X$, let n_x denote the number of neighbours of x. Since all connections have bounded range, we have that $n_x < \infty$. Now, if $n_x \leq 1$, the proposition is trivially true. Let us consider the case $n_x > 1$, and denote by x_1 the node connected to x that satisfies

$$Pl(|x_1 - x|) \leq Pl(|x_i - x|), \quad \text{for all } i = 2, \ldots, n_x. \tag{2.76}$$

Since x_1 is connected to x we have that

$$\frac{Pl(|x_1 - x|)}{N + \gamma \sum_{i=2}^{\infty} Pl(|x_i - x|)} \geq T. \tag{2.77}$$

Taking (2.76) into account we have

$$Pl(|x_1 - x|) \geq TN + T\gamma \sum_{i=2}^{\infty} Pl(|x_i - x|)$$

$$\geq TN + T\gamma(n_x - 1)Pl(|x_1 - x|) + T\gamma \sum_{i=n_x+1}^{\infty} Pl(|x_i - x|)$$

$$\geq T\gamma(n_x - 1)Pl(|x_1 - x|), \tag{2.78}$$

from which we deduce that

$$n_x \leq 1 + \frac{1}{T\gamma}. \tag{2.79}$$

$\square$

The next two theorems characterise the phase transition in the interference model. The first one shows that percolation occurs beyond the critical density value of the boolean model, by taking γ sufficiently small.

Theorem 2.7.4 *Let λ_c be the critical node density when $\gamma = 0$. For any node density $\lambda > \lambda_c$, there exists $\gamma^*(\lambda) > 0$ such that for $\gamma \leq \gamma^*(\lambda)$, the interference model percolates.*

The second theorem shows that for any fixed γ, increasing the density of the Poisson point process always leads to a disconnected network.

Theorem 2.7.5 *For $\lambda \to \infty$ we have that*

$$\gamma^*(\lambda) = O\left(\frac{1}{\lambda}\right). \tag{2.80}$$

The bounds on the supercritical region expressed by Theorems 2.7.4, 2.7.5, and Proposition 2.7.3, are visualised in Figure 2.17, which can now be compared with the numerical results depicted in Figure 2.16.

The proof of Theorem 2.7.4 is divided into different steps. The main strategy is to couple the model with a discrete bond percolation model on the grid. By doing so, we end up with a dependent discrete model, such that the existence of an infinite connected component in the bond percolation model implies the existence of an infinite connected component in the original graph. Although the edges of the discrete model are not finite range dependent, we show that the probability of not having a collection of n edges in the random grid decreases exponentially as q^n, where q can be made arbitrarily small by an appropriate choice of the parameters, and therefore the existence of an infinite connected component follows from a Peierls argument such as the one in Theorem 2.2.5.

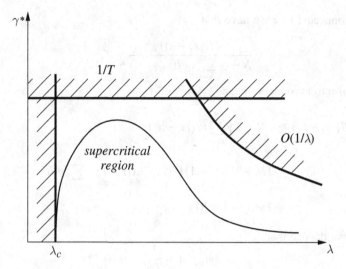

Fig. 2.17   Illustration of the bounds on the supercritical region.

We describe the construction of the discrete model first, then we prove percolation on the discrete grid, and finally we obtain the final result by coupling the interference model and the discrete process.

2.7.1 Mapping on a square lattice

If we let $\gamma = 0$, the interference model coincides with a Poisson boolean model of radius r_b given by

$$2r_b = l^{-1}\left(\frac{TN}{P}\right). \tag{2.81}$$

Since l is continuous, strictly monotone and larger than TN/P at the origin, we have that $l^{-1}(TN/P)$ exists.

We consider next a supercritical boolean model of radius r_b, where the node density λ is higher than the critical value λ_c. By rescaling the model, we can establish that the critical radius for a fixed density $\lambda > \lambda_c$ is

$$r^*(\lambda) = \sqrt{\frac{\lambda_c}{\lambda}} r_b < r_b. \tag{2.82}$$

Therefore, a boolean model with density λ and radius r satisfying $r^*(\lambda) < r < r_b$, is still supercritical.

We map this latter model into a discrete percolation model as follows. We denote by G_d the two-dimensional square lattice with nearest neighbour vertices spaced by distance $d > 0$. Choosing an arbitrary lattice point as the origin, for each horizontal edge $a \in G_d$, we denote by z_a the point in the middle of the edge, with coordinates (x_a, y_a),

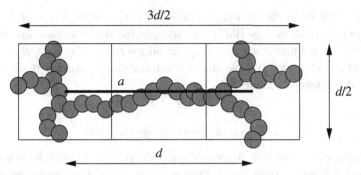

Fig. 2.18 A horizontal edge a that fulfills the two conditions for having $A_a = 1$.

and introduce the random variable A_a that takes value one if the following two events (illustrated in Figure 2.18) occur, and zero otherwise:

(i) the rectangle $[x_a - 3d/4, x_a + 3d/4] \times [y_a - d/4, y_a + d/4]$ is crossed from left to right by a component of the boolean model, and

(ii) both squares $[x_a - 3d/4, x_a - d/4] \times [y_a - d/4, y_a + d/4]$ and $[x_a + d/4, x_a + 3d/4] \times [y_a - d/4, y_a + d/4]$ are crossed from top to bottom by a component of the boolean model.

We define A_a similarly for vertical edges, by rotating the above conditions by 90°. By Theorem 2.7.1, the probability that $A_a = 1$ can be made as large as we like by choosing d large. Note that variables A_a are not independent in general. However, if a and b are not adjacent, then A_a and A_b are independent: these variables thus define a 1-dependent bond percolation process.

We now define a shifted version $\tilde{l}$ of the function l as follows:

$$
\tilde{l}(x) = \begin{cases} l(0) & x \le \dfrac{\sqrt{10}d}{4}, \\ l\left(x - \dfrac{\sqrt{10}d}{4}\right) & x > \dfrac{\sqrt{10}d}{4}. \end{cases} \tag{2.83}
$$

We also define the *shot-noise* processes I and $\tilde{I}$ at any $z \in \mathbb{R}^2$ by taking the following infinite sums over all Poisson points of X:

$$
I(z) = \sum_{x \in X} l(|z - x|) \tag{2.84}
$$

and

$$
\tilde{I}(z) = \sum_{x \in X} \tilde{l}(|z - x|), \tag{2.85}
$$

Note that the shot-noises are random variables, since they depend on the random position of the points of X.

We define now a second indicator random variable B_a that takes value one if the value of the shot-noise $\tilde{I}(z_a)$ does not exceed a certain threshold $M > 0$. As the distance between any point z inside the rectangle $R(z_a) = [x_a - 3d/4, x_a + 3d/4] \times [y_a - d/4, y_a + d/4]$ and its centre z_a is at most $\sqrt{10}d/4$, the triangle inequality implies that $|z_a - x| \leq \sqrt{10}d/4 + |z - x|$, and thus that $I(z) \leq \tilde{I}(z_a)$ for all $z \in R(z_a)$. Therefore, $B_a = 1$ implies that $I(z) \leq M$ for all $z \in R(z_a)$. Note also that in this case the variables B_a do not have a finite range dependency structure.

2.7.2 Percolation on the square lattice

For any edge a of G_d, we call the edge *good* if the product $C_a = A_a B_a$ is one, that is if both of the following events occur: there exist crossings in the rectangle $R(z_a)$ *and* the shot-noise is bounded by M for all points inside $R(z_a)$. If an edge a is not good we call it *bad*. We want to show that for appropriate choice of the parameters M and d, there exists an infinite connected component of good edges at the origin, with positive probability.

To do this, all we need is an exponential bound on the probability of a collection of n bad edges. Then, percolation follows from the standard Peierls argument. Most of the difficulty of obtaining this resides in the infinite range dependencies introduced by the random variables B_i. Fortunately, a careful application of Campbell's Theorem will take care of this, as shown below. In the following, to keep the notation simple, we write $A_{a_i} = A_i$, $B_{a_i} = B_i$ and $C_{a_i} = C_i$, for $i = 1, \ldots, n$.

Lemma 2.7.6 *Let $\{a_i\}_{i=1}^n$ be a collection of n distinct edges, and $\{C_i\}_{i=1}^n$ the random variables associated with them. Then there exists $q_C < 1$, independent of the particular collection, such that*

$$P(C_1 = 0, C_2 = 0, \ldots, C_n = 0) \leq q_C^n. \tag{2.86}$$

Furthermore, for any $\epsilon > 0$, one can choose d and M so that $q_C \leq \epsilon$.

It should be clear that with Lemma 2.7.6, the existence of an unbounded component in the dependent bond percolation model immediately follows from a Peierls argument as described in the proof of Theorem 2.2.5.

We prove first the exponential bound separately for A_i and B_i and then combine the two results to prove Lemma 2.7.6 and thus percolation in the dependent edge model.

Lemma 2.7.7 *Let $\{a_i\}_{i=1}^n$ be a collection of n distinct edges, and let $\{A_i\}_{i=1}^n$ be the random variables associated with them. Then there exists $q_A < 1$, independent of the particular collection, such that*

$$P(A_1 = 0, A_2 = 0, \ldots, A_n = 0) \leq q_A^n. \tag{2.87}$$

Furthermore, for any $\epsilon > 0$, one can choose d large enough so that $q_A \leq \epsilon$.

Proof We can easily prove this lemma following the same argument as in Theorem 2.3.1, adapted to the bond percolation case. We observe that for our 1-dependent bond percolation model, it is always possible to find a subset of indices $\{k_j\}_{j=1}^m$ with

$1 \leq k_j \leq n$ for each j, such that the variables $\{A_{k_j}\}_{j=1}^m$ are independent and $m \geq n/4$. Therefore we have

$$P(A_1 = 0, A_2 = 0, \ldots, A_n = 0) \leq P\big(A_{k_1} = 0, A_{k_2} = 0, \ldots, A_{k_m} = 0\big)$$
$$= P(A_1 = 0)^m$$
$$\leq P(A_1 = 0)^{\frac{n}{4}}$$
$$\equiv q_A^n. \tag{2.88}$$

Furthermore, since $q_A = P(A_1 = 0)^{1/4}$, it follows from Theorem 2.7.1 that q_A tends to zero when d tends to infinity. $\qquad\square$

Lemma 2.7.8 *Let $\{a_i\}_{i=1}^n$ be a collection of n distinct edges, and $\{B_i\}_{i=1}^n$ the random variables associated with them. Then there exists $q_B < 1$, independent of the particular collection, such that*

$$P(B_1 = 0, B_2 = 0, \ldots, B_n = 0) \leq q_B^n. \tag{2.89}$$

Furthermore, for any $\epsilon > 0$ and fixed d, one can choose M large enough so that $q_B \leq \epsilon$.

Proof The proof of this lemma is more involved because in this case dependencies are not of finite range. We will find an exponential bound by applying Campbell's Theorem. To simplify notation, we denote by z_i the centre z_{a_i} of edge a_i. By Markov's inequality (see Appendix A.4.1), we have for any $s \geq 0$,

$$P(B_1 = 0, B_2 = 0, \ldots, B_n = 0) \leq P\big(\tilde{I}(z_1) > M, \tilde{I}(z_2) > M, \ldots, \tilde{I}(z_n) > M\big)$$
$$\leq P\left(\sum_{i=1}^n \tilde{I}(z_i) > nM\right)$$
$$\leq e^{-snM} E\left(e^{s \sum_{i=1}^n \tilde{I}(z_i)}\right). \tag{2.90}$$

We use Campbell's Theorem 2.7.2 applied to the function

$$f(x) = \sum_{i=1}^n \tilde{l}(|x - z_i|). \tag{2.91}$$

Note that this is possible because the integrability condition on the attenuation function can be easily extended to $\tilde{l}$, that is

$$\int_y^\infty x \tilde{l}(x) dx < \infty \text{ for some } y > 0, \tag{2.92}$$

$$\tilde{l}(x) \leq 1 \text{ for all } x \in \mathbb{R}^+, \tag{2.93}$$

and (2.92), (2.93) immediately imply (2.67). Accordingly, we obtain

$$E\left(e^{s \sum_{i=1}^n \tilde{I}(z_i)}\right) = \exp\left(\lambda \int_{\mathbb{R}^2} (e^{s \sum_{i=1}^n \tilde{l}(|x - z_i|)} - 1) dx\right). \tag{2.94}$$

We need to estimate the exponent $s \sum_{i=1}^n \tilde{l}(|x - z_i|)$. As $\{z_i\}$ are centres of edges, they are located on a square lattice tilted by 45°, with edge length $d/\sqrt{2}$; see Figure 2.19. So, if

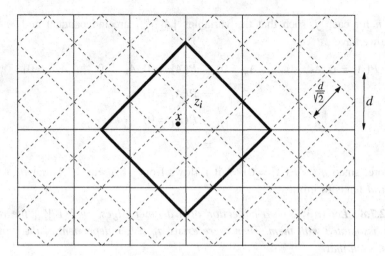

Fig. 2.19   The tilted lattice defined by points $\{z_i\}$.

we consider the square in which x is located, the contribution to $\sum_{i=1}^{n} \tilde{l}(|x - z_i|)$ coming from the four corners of this square is at most equal to four, since $\tilde{l}(x) \leq 1$. Around this square, there are 12 nodes, each located at distance at least $d/\sqrt{2}$ from x. Further away, there are 20 other nodes at distance at least $2d/\sqrt{2}$, and so on. Consequently,

$$\sum_{i=1}^{n} \tilde{l}(|x - z_i|) \leq \sum_{i=1}^{\infty} \tilde{l}(|x - z_i|)$$

$$\leq 4 + \sum_{k=1}^{\infty} (4 + 8k)\tilde{l}\left(\frac{kd}{\sqrt{2}}\right) \equiv K. \tag{2.95}$$

Using the integral criterion and (2.92), we conclude that the sum converges and thus $K < \infty$.

The computation made above holds for any $s \geq 0$. We now take $s = 1/K$, so that $s \sum_{i=1}^{n} \tilde{l}(|x - z_i|) \leq 1$, for all x. Furthermore, since $e^x - 1 < 2x$ for all $x \leq 1$, we have

$$e^{s \sum_{i=1}^{n} \tilde{l}(|x - z_i|)} - 1 < 2s \sum_{i=1}^{n} \tilde{l}(|x - z_i|) = \frac{2}{K} \sum_{i=1}^{n} \tilde{l}(|x - z_i|). \tag{2.96}$$

Substituting (2.96) in (2.94), we obtain

$$E\left(e^{\sum_{i=1}^{n} \tilde{l}(z_i)/K}\right) \leq \exp\left(\lambda \int_{\mathbb{R}^2} \frac{2}{K} \sum_{i=1}^{n} \tilde{l}(|x - z_i|)dx\right)$$

$$= \exp\left(\frac{2n\lambda}{K} \int_{\mathbb{R}^2} \tilde{l}(|x|)dx\right)$$

$$= \left[\exp\left(\frac{2\lambda}{K} \int_{\mathbb{R}^2} \tilde{l}(|x|)dx\right)\right]^n. \tag{2.97}$$

Putting things together, we have that

$$
\begin{aligned}
P\big(\tilde{I}(z_1) &> M, \tilde{I}(z_2) > M, \ldots, \tilde{I}(z_n) > M\big) \\
&\leq e^{-snM} E\Big(e^{s\sum_{i=1}^{n} \tilde{I}(z_i)}\Big) \\
&\leq e^{-nM/K} \left[\exp\left(\frac{2\lambda}{K}\int_{\mathbb{R}^2} \tilde{l}(|x|)dx\right)\right]^n \\
&= q_B^n,
\end{aligned}
\tag{2.98}
$$

where q_B is defined as

$$
q_B \equiv \exp\left(\frac{2\lambda}{K}\int_{\mathbb{R}^2} \tilde{l}(|x|)dx - \frac{M}{K}\right).
\tag{2.99}
$$

Furthermore, it is easy to observe that this expression tends to zero when M tends to infinity. $\qquad\square$

We are now ready to combine the two results above and prove Lemma 2.7.6.

Proof of Lemma 2.7.6 For convenience, we introduce the following notation for the indicator random variables A_i and B_i. $\bar{A}_i = 1 - A_i$ and $\bar{B}_i = 1 - B_i$.

First observe that

$$
1 - C_i = 1 - A_i B_i \leq (1 - A_i) + (1 - B_i) = \bar{A}_i + \bar{B}_i.
\tag{2.100}
$$

Let us denote by $p(n)$ the probability that we want to bound, and let $(k_i)_{i=1}^n$ be a binary sequence (i.e., $k_i = 0$ or 1) of length n. We denote by $\mathcal{K}$ the set of the 2^n such sequences. Then we can write

$$
\begin{aligned}
p(n) &= P(C_1 = 0, C_2 = 0, \ldots, C_n = 0) \\
&= E((1 - C_1)(1 - C_2)\cdots(1 - C_n)) \\
&\leq E\big((\bar{A}_1 + \bar{B}_1)(\bar{A}_2 + \bar{B}_2)\cdots(\bar{A}_n + \bar{B}_n)\big) \\
&= \sum_{(k_i)\in\mathcal{K}} E\left(\prod_{i:k_i=0} \bar{A}_i \prod_{i:k_i=1} \bar{B}_i\right) \\
&\leq \sum_{(k_i)\in\mathcal{K}} \sqrt{E\left(\prod_{i:k_i=0} (\bar{A}_i)^2\right) E\left(\prod_{i:k_i=1} (\bar{B}_i)^2\right)} \\
&= \sum_{(k_i)\in\mathcal{K}} \sqrt{E\left(\prod_{i:k_i=0} \bar{A}_i\right) E\left(\prod_{i:k_i=1} \bar{B}_i\right)},
\end{aligned}
\tag{2.101}
$$

where the last inequality follows from the Cauchy–Schwarz inequality (see Appendix A.5), and the last equality from the observation that $(\bar{A}_i)^2 = \bar{A}_i$ and $(\bar{B}_i)^2 = \bar{B}_i$. Applying

Propositions 2.7.7 and 2.7.8, we can bound each expectation in the sum. We have thus

$$p(n) \leq \sum_{(k_i) \in \mathcal{K}} \sqrt{\prod_{i:k_i=0} q_A \prod_{i:k_i=1} q_B}$$

$$= \sum_{(k_i) \in \mathcal{K}} \prod_{i:k_i=0} \sqrt{q_A} \prod_{i:k_i=1} \sqrt{q_B}$$

$$= \left(\sqrt{q_A} + \sqrt{q_B}\right)^n$$

$$\equiv q_C^n. \tag{2.102}$$

Choosing first d and then M appropriately, we can make q_C smaller than any given ϵ. $\square$

2.7.3 Percolation of the interference model

We can now finalise the proof of percolation in the original interference model, coupling this model with the dependent percolation model.

Proof of Theorem 2.7.4 We want to show that percolation in the discrete model implies percolation in the interference model, with appropriate γ. The value of γ that we shall choose to make the model percolate depends on λ through the parameter M, and on the attenuation function.

We start by noticing that if $B_a = 1$, the interference level in the rectangle $R(z_a)$ is at most equal to M. Therefore, for two nodes x_i and x_j in $R(z_a)$ such that $|x_i - x_j| \leq 2r$, we have

$$\frac{Pl(|x_i - x_j|)}{N + \gamma \sum_{k \neq i,j} Pl(|x_k - x_j|)} \geq \frac{Pl(|x_i - x_j|)}{N + \gamma PM}$$

$$\geq \frac{Pl(2r)}{N + \gamma PM}. \tag{2.103}$$

As $r < r_b$ and as l is strictly decreasing, we pick

$$\gamma = \frac{N}{PM} \left(\frac{l(2r)}{l(2r_b)} - 1\right) > 0, \tag{2.104}$$

yielding

$$\frac{Pl(2r)}{N + \gamma PM} = \frac{Pl(2r_b)}{N} = T. \tag{2.105}$$

Therefore, there exists a positive value of γ such that any two nodes separated by a distance less than r are connected in the interference model. This means that in the rectangle $R(z_a)$ all connections of the boolean model of parameters λ and r also exist in the interference model.

Finally, if $A_a = 1$, there exist crossings along edge a, as shown in Figure 2.18. These crossings are designed such that if for two adjacent edges a and b, $A_a = 1$ and $A_b = 1$, the crossings overlap, and they all belong to the same connected component; see Figure 2.20.

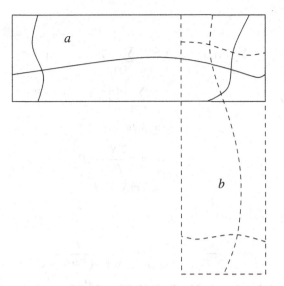

Fig. 2.20   Two adjacent edges a (plain) and b (dashed) with $A_a = 1$ and $A_b = 1$. The crossings overlap, and form a unique connected component.

Thus, an infinite cluster of such edges implies an infinite cluster in the boolean model of radius r and density λ. Since all edges a of the infinite cluster of the discrete model are such that $A_a = 1$ and $B_a = 1$, this means that the crossings also exist in the interference model, and thus form an infinite connected component.   □

2.7.4 Bound on the percolation region

We now want to give a proof of Theorem 2.7.5. Notice that this is an asymptotic statement that implies there is no percolation for λ large enough, while in the proof of Theorem 2.7.4 we have first fixed λ and then chosen a corresponding value of $\gamma(\lambda)$ that allows percolation. In order to prove Theorem 2.7.5, we start by showing a preliminary technical lemma. Consider an infinite square grid G, similar to the previous one, but with edge length $\delta/2$ instead of d and let s be an arbitrary lattice cell of G.

Lemma 2.7.9   *If there are more than*

$$m = \frac{(1+2T\gamma)P}{\gamma NT^2} \tag{2.106}$$

nodes inside s, then all nodes in s are isolated.

Proof   Let $x_i \in X$ be a node located inside s, and let x_j be any other node of X. Clearly, as $l(\cdot)$ is bounded from above by one, we have that $Pl(|x_j - x_i|) \leq P$. Also recall that $l(0) > TN/P$ and that $l(\cdot)$ is a continuous and decreasing function. It follows that

$$\sum_{k \neq i,j} Pl(|x_k - x_i|) \geq \sum_{x_k \in s, k \neq i,j} Pl(|x_k - x_i|)$$

$$\geq \sum_{x_k \in s} Pl(|x_k - x_i|) - 2P$$

$$\geq mPl(0) - 2P$$

$$\geq mP\frac{TN}{P} - 2P$$

$$= TmN - 2P. \tag{2.107}$$

Therefore we have

$$\frac{Pl(|x_j - x_i|)}{N + \gamma \sum Pl(|x_k - x_i|)} \leq \frac{P}{N + \gamma(TmN - 2P)}$$

$$\leq \frac{P}{\gamma(TmN - 2P)} \tag{2.108}$$

The above expression is clearly smaller than T when

$$m > \frac{(1 + 2T\gamma)P}{\gamma NT^2}, \tag{2.109}$$

which implies that node x_i is isolated. $\qquad\square$

Proof of Theorem 2.7.5 We now consider a site percolation model on G, by declaring each box of the grid *open* if it contains at most

$$2m = 2\frac{(1 + 2T\gamma)P}{\gamma NT^2} \tag{2.110}$$

nodes, *closed* otherwise. We call boxes that share at least a point *neighbours* and boxes that share a side *adjacent*. Note that with these definitions every finite cluster of neighbouring open sites is surrounded by a circuit of adjacent closed sites and vice versa; you can look back at Figure 2.5 for an illustration. Furthermore, it is clear that each site is open or closed independently of the others and that each closed site contains only isolated nodes. Let us denote by $|s|$ the number of Poisson points located inside a site s. Since the area of a site is $\delta^2/4$, by Chebyshev's inequality (see Appendix A.4.2), we have that for any $\epsilon > 0$,

$$P\left(|s| \leq \frac{(1 - \epsilon)\lambda\delta^2}{4}\right) \leq \frac{4}{\epsilon^2\lambda\delta^2}. \tag{2.111}$$

Next we choose $\gamma(\lambda)$ be such that

$$2m = \frac{(1 - \epsilon)\lambda\delta^2}{4}, \tag{2.112}$$

that is, by (2.110) we let γ be such that

$$\gamma = \frac{4P}{T^2 N(1-\epsilon)\delta^2\lambda - 8TP}. \qquad (2.113)$$

With the above choice we have that as $\lambda \to \infty$, $\gamma = O(1/\lambda)$, and also by (2.111), (2.112),

$$P(|s| \le 2m) = P\left(|s| \le \frac{(1-\epsilon)\lambda\delta^2}{4}\right) \le \frac{4}{\epsilon^2\lambda\delta^2} \to 0. \qquad (2.114)$$

This shows that for λ large enough the discrete site model defined by neighbouring sites does not percolate. We now have to prove that in this case also the original continuous interference model is subcritical. It then immediately follows that $\gamma^*(\lambda) = O(1/\lambda)$, as $\lambda \to \infty$.

We start by noticing that since the discrete site percolation model defined by neighbouring boxes is subcritical, the origin is a.s. surrounded by a circuit of adjacent closed boxes. By Lemma 2.7.9, when a site is closed it contains only Poisson points that are isolated in the interference model. Therefore, the origin is a.s. surrounded by a chain of boxes with no edge incident inside them. To make sure that the origin belongs to a finite cluster, we have to prove that no link can cross this chain.

Let us consider two nodes x_i and x_j, such that x_i is in the interior of the chain, and x_j is located outside the chain. As the chain of closed sites passes between these nodes, the distance d between them is larger than $\delta/2$; see Figure 2.21.

We consider two cases. First, we assume that $\delta/2 < d < \delta$. In this case let D_1 be the disc of radius δ centred at x_i and D_2 be the disc of radius δ centred at x_j, as depicted in Figure 2.21. Let Q be a square of the chain that has a non-empty intersection with the segment joining x_i and x_j. Note that the shortest distance between this segment and $\mathbb{R}^2 \setminus (D_1 \cup D_2)$ is

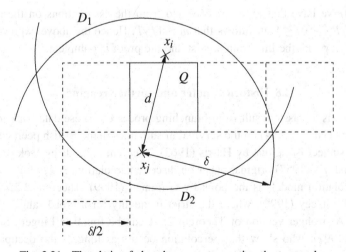

Fig. 2.21   The chain of closed squares separating the two nodes.

$$\sqrt{\delta^2 - \frac{d^2}{4}} = \frac{\sqrt{3}}{2}\delta. \qquad (2.115)$$

As the diagonal of Q has length $\delta\sqrt{2}/2$, it follows that $Q \subset D_1 \cup D_2$. We now let

$$N_1 = |Q \cap (D_1 \setminus D_2)|,$$
$$N_2 = |(Q \cap (D_2 \setminus D_1)|,$$
$$N_3 = |Q \cap D_1 \cap D_2|, \qquad (2.116)$$

where we have indicated by $|\cdot|$ the number of Poisson points inside a given region. Since Q is a closed square, we have that $N_1 + N_2 + N_3 \geq 2m$. This implies that either $N_1 + N_3 \geq m$, or $N_2 + N_3 \geq m$. Let us assume that the first inequality holds. There are thus at least m nodes located inside D_1. Since $l(0) > TN/P$, by continuity of $l(\cdot)$ we can choose δ small enough so that $l(\delta) > TN/P$. As D_1 has radius δ, the SNIR at x_i can now be at most

$$\frac{P}{N + \gamma \frac{PmTN}{P}} < \frac{P}{\gamma mTN} < T, \qquad (2.117)$$

where the last inequality follows by exploiting (2.110). We conclude that no link between x_i and x_j exists. The same is true if $N_2 + N_3 \geq m$.

Let us now address the case where $d > \delta$ (the case $d = \delta$ has probability zero). In this case, we draw the same discs D_1 and D_2, but with radius d. There exists at least one square Q of the chain such that $Q \subset D_1 \cup D_2$. We define N_1, N_2 and N_3 in the same way as above. Thus, either $N_1 + N_3 \geq m$ or $N_2 + N_3 \geq m$. Let us assume without loss of generality that $N_1 + N_3 \geq m$. This implies that there are at least m nodes inside D_1. Node x_j is by construction on the border of D_1. Therefore, all nodes inside D_1 are closer to x_i than node x_j. Since $l(\cdot)$ is decreasing, the SNIR at i is at most

$$\frac{Pl(d)}{N + \gamma Pml(d)} \leq \frac{Pl(d)}{\gamma Pml(d)} \leq \frac{1}{\gamma m}. \qquad (2.118)$$

From (2.110) we have that $m > P/\gamma T^2 N$ and from the assumptions on the attenuation function $TN/P < l(0) \leq 1$, it follows that $m > 1/\gamma T$. Hence the above expression is less than T, implying that the link cannot exist and the proof is complete. □

2.8 Historical notes and further reading

Theorem 2.1.1 is a classic result from branching processes. These date back to the work of Watson and Galton (1874) on the survival of surnames in the British peerage. A classic book on the subject is the one by Harris (1963). Theorem 2.2.5 dates back to Broadbent and Hammersley's (1957) original paper on discrete percolation. A general reference for discrete percolation models is the book by Grimmett (1999). Theorem 2.2.7 appears in Grimmett and Stacey (1998), where the strict inequality for a broad range of graphs is also shown. A stronger version of Theorem 2.3.1 can be found in Liggett, Schonmann, and Stacey (1997), who show that percolation occurs as long as the occupation probability of any site (edge), conditioned to the states of all sites (edges) outside a finite neighbourhood of it, can be made sufficiently high. Theorem 2.4.1 is by Häggström and Meester (1996). Theorem 2.5.1 is by Penrose (1991). Theorems 2.5.2, 2.5.4, 2.5.5 are by Franceschetti, Booth *et al.* (2005). Theorem 2.5.4 was also discovered independently by Balister, Bollobás, and Walters (2004). Similar results, for a different spreading transformation, appear in Penrose (1993), and Meester, Penrose, and Sarkar (1997). Theorem 2.5.6 is by Penrose (1991), while compression results for the boolean model appear in Alexander (1991). Theorem 2.6.1 dates back to Gilbert (1961). A general reference for continuum models is the book by Meester and Roy (1996). The treatment of the signal to noise plus interference model follows Dousse, Baccelli and Thiran (2005) and Dousse, Franceschetti, Macris, *et al.* (2006).

Exercises

2.1 Derive upper and lower bounds for the critical value p_c for site percolation on the square lattice following the outline of the proof of Theorem 2.2.5.

2.2 Find graphs with bond percolation critical value equal to zero and one respectively.

2.3 Find a graph with $0 < p_c^{bond} = p_c^{site} < 1$.

2.4 Prove the following statement (first convince yourself that there is actually something to prove at all!): in percolation on the two-dimensional integer lattice the origin is in an infinite component if and only if there is an open path from the origin to infinity.

2.5 Consider bond percolation on the one-dimensional line, where each edge is deleted with probability $p = 1/2$. Consider a segment of n edges. What is the probability that the vertices at the two sides of the segment are connected ? What happens if n increases? Consider now bond percolation on the square lattice again with $p = 1/2$. Consider a square with n vertices at each side. What is the probability that there exists a path connecting the left side with the right side?

2.6 In the proof of Theorem 2.2.7 we have compared the dynamic marking procedure with site percolation, and argued that if $P(|C| = \infty) > 0$ in the site percolation model, then this is also true for the tree obtained using the marking procedure. Note that if we could argue the same for bond percolation, we could prove that $p_c^{site} = p_c^{bond}$. Where does the argument for bond percolation fail?

2.7 Prove that in the random connection model, $\theta(\lambda)$ is non-decreasing in λ. In order to do this you need to use that given a realisation of a Poisson process of density λ on the plane, and deleting each point independently from this realisation with probability $(1 - p)$, you obtain a realisation of a Poisson process of density $p\lambda$.

2.8 Consider a boolean model in one dimension, that is, on a line (balls are now just intervals). Suppose that we put intervals of fixed length around each point of the point process. Explain why the critical density is now equal to $\lambda_c = \infty$.

2.9   Consider a boolean model in one dimension, that is, on a line (balls are now just intervals). Suppose that we put intervals of *random* length around each point. All lengths are identically distributed and independent of each other, and we let R denote a random variable with this length-distribution. Prove that when $E(R) = \infty$, the critical density is equal to $\lambda_c = 0$.

2.10 Consider the continuum percolation model on the full plane, where each point of a Poisson point process connects itself to its k nearest neighbours. We denote by $f(k)$ the probability that the point at the origin (we assume we have added a point at the origin) is contained in an infinite cluster. Show that if $f(k) > 0$, then $f(k+1) > f(k)$ (strict inequality).

2.11 Prove the claim made towards the end of the proof of Theorem 2.5.2.

2.12 In the boolean model, it is believed that the critical value of the average number of connections per node to percolate is $\xi_c \approx 4.512$. By using scaling relations, compare this value with the lower bound obtained in the proof of Proposition 2.5.3. Why does this lower bound apply to the boolean model?

2.13 Provide an upper bound for the critical density required for percolation of the boolean model.

2.14 Prove Theorem 2.3.1 for the bond percolation case and compare the exponential bound for the k-dependent model with the one in the proof of Lemma 2.7.7.

2.15 Prove Theorem 2.2.4.

2.16 Give all details of the proof of Theorem 2.3.1. In particular, can you give explicit values of $p_1(k)$ and $p_2(k)$?

2.17 Show that the percolation function in the random connection model is non-decreasing.

2.18 Can you improve inequality (2.29)?

2.19 Complete the proof of Theorem 2.7.2.

2.20 Explain why we need to choose d before M at the end of the proof of Proposition 2.7.6.

2.21 Explain why property (ii) of the attenuation function on page 53 implies convergence of the series in (2.75).

3

Connectivity of finite networks

One of the motivations to study random networks on the infinite plane has been the possibility of observing sharp transitions in their behaviour. We now discuss the asymptotic behaviour of sequences of finite random networks that grow larger in size. Of course, one expects that the sharp transitions that we observe on the infinite plane are a good indication of the limiting behaviour of such sequences, and we shall see to what extent this intuition is correct and can be made rigorous.

In general, asymptotic properties of networks are of interest because real systems are of finite size and one wants to discover the correct *scaling laws* that govern their behaviour. This means discovering how the system is likely to behave as its size increases.

We point out that there are two equivalent scalings that produce networks of a growing number of nodes: one can either keep the area where the network is observed fixed, and increase the density of the nodes to infinity; or one can keep the density constant and increase the area of interest to infinity. Although the two cases above can describe different practical scenarios, by appropriate scaling of the distance lengths, they can be viewed as the same network realisation, so that all results given in this chapter apply to both scenarios.

3.1 Preliminaries: modes of convergence and Poisson approximation

We make frequent use of a powerful tool, the *Chen–Stein method*, to estimate convergence to a Poisson distribution. This method is named after work of Chen (1975) and Stein (1978) and is the subject of the monograph by Barbour, Holst and Janson (1992). We have already seen in Chapter 1 how a Poisson distribution naturally arises as the limiting distribution of the sum of n independent, low probability, indicator random variables. The idea behind the Chen–Stein method is that this situation generalises to dependent, low probability random variables, as long as dependencies are negligible as n tends to infinity, broadly speaking. To set things up correctly, we first define a *distance* between two probability distributions and various *modes of convergence* of sequences of random variables.

Definition 3.1.1 *The total variation distance between two probability distributions p and q on $\mathbb{N}$ is defined by*

$$d_{TV}(p, q) = \sup\{|p(A) - q(A)| : A \subset \mathbb{N}\}. \tag{3.1}$$

Definition 3.1.2 *A sequence X_n of random variables converges almost surely to X if*

$$P\left(\lim_{n\to\infty} X_n = X\right) = 1. \tag{3.2}$$

A sequence X_n of random variables converges in probability to X if for all $\epsilon > 0$,

$$\lim_{n\to\infty} P(|X_n - X| > \epsilon) = 0. \tag{3.3}$$

Finally, if X_n takes values in $\mathbb{N}$ for all n, then we say that X_n converges in distribution to X if for all $k \in \mathbb{N}$ we have

$$\lim_{n\to\infty} P(X_n \le k) = P(X \le k). \tag{3.4}$$

It is clear that the strongest mode of convergence is the almost sure convergence, which implies convergence in probability, which in turn implies convergence in distribution. The latter is sometimes referred to as *weak convergence*. Also note that weak convergence is equivalent to having $d_{TV}(X_n, X)$ tending to zero as $n \to \infty$, where we have identified the random variables X_n and X with their distributions.

We now introduce some bounds on the total variation distance between the Poisson distribution of parameter λ on the one hand, and the distribution of the sum of n dependent indicator random variables with expectations p_α on the other hand. We refer the reader to the book by Barbour, Holst and Janson (1992) for a complete treatment of Chen–Stein bounds of this kind. One bound that we use holds when the indicator variables are increasing functions of independent random variables.

We introduce the following notation. Let $\mathcal{I}$ be an arbitrary index set, and for $\alpha \in \mathcal{I}$, let I_α be an indicator random variable with expectation $E(I_\alpha) = p_\alpha$. We define

$$\lambda = \sum_{\alpha \in \mathcal{I}} p_\alpha \tag{3.5}$$

and assume that $\lambda < \infty$. Let $W = \sum_{\alpha \in \mathcal{I}} I_\alpha$, and note that $E(W) = \lambda$. Finally, $Po(\lambda)$ denotes a Poisson random variable with parameter λ.

Theorem 3.1.3 *If the I_α are increasing functions of independent random variables $X_1, \ldots, X_k$, then we have*

$$d_{TV}(W, Po(\lambda)) \le \frac{1 - e^{-\lambda}}{\lambda}\left(\text{Var } W - \lambda + 2\sum_{\alpha \in \mathcal{I}} p_\alpha^2\right). \tag{3.6}$$

Another bound we use appears in Arratia *et al.* (1989), and makes use of the notion of *neighbourhood of dependence*, as defined below.

Definition 3.1.4 *For each $\alpha \in \mathcal{I}$, $B_\alpha \subset \mathcal{I}$ is a neighbourhood of dependence for α, if I_α is independent of all indices I_β, for $\beta \notin B_\alpha$.*

Theorem 3.1.5 *Let B_α be a neighbourhood of dependence for $\alpha \in \mathcal{I}$. Let*

$$b_1 \equiv \sum_{\alpha \in \mathcal{I}} \sum_{\beta \in B_\alpha} E(I_\alpha) E(I_\beta),$$

$$b_2 \equiv \sum_{\alpha \in \mathcal{I}} \sum_{\beta \in B_\alpha, \beta \neq \alpha} E(I_\alpha I_\beta). \tag{3.7}$$

It is the case that

$$d_{TV}(W, Po(\lambda)) \leq 2(b_1 + b_2). \tag{3.8}$$

The application we make of the bounds above is by considering indicator random variables of events in random networks whose probability decays with n. In this case the bounds converge to zero and the sum of the indicators converges in distribution to the Poisson distribution with parameter $\lambda = E(W)$.

3.2 The random grid

We start by looking at the random grid. We are interested in discovering *scaling laws* of sequences of finite networks contained in a box of size $n \times n$. These scaling laws are events that occur asymptotically almost surely (a.a.s.), meaning with probability tending to one as as $n \to \infty$. We also use the terminology with high probability (w.h.p.), to mean the same thing.

We have seen in Chapter 2 that on the infinite lattice, an unbounded connected component of vertices forms when the edge (site) probability exceeds a critical threshold value p_c. By looking at the same model restricted to a finite box, we might reasonably expect that above p_c the percolation probability $\theta(p) > 0$ roughly represents the average number of vertices that are connected inside a finite box. It turns out that this is indeed the case in the limit for the box size that tends to infinity. On the other hand, we also expect that in order to obtain a *fully* connected network inside the finite box, the value of p must be close to one, and we shall give the precise rate by which p must approach one.

We start by looking at the fraction of connected vertices above criticality, where similar results hold for site, bond, and continuum percolation models.

3.2.1 Almost connectivity

We call G_n the $n \times n$ random grid with edge (site) probability p.

Definition 3.2.1 *For any $\alpha \in (0, 1)$, G_n is said to be α-almost connected if it contains a connected component of at least αn^2 vertices.*

Theorem 3.2.2 *Let*

$$p_\alpha = \inf\{p; \theta(p) > \alpha\}. \tag{3.9}$$

For any $\alpha \in (0, 1)$, we have that if $p > p_\alpha$, then G_n is α-almost connected a.a.s., while for $p < p_\alpha$ it is not.

The theorem above states that the percolation function asymptotically corresponds to the fraction of connected nodes in G_n. A corresponding theorem can also be stated for the boolean model, by replacing $\theta(p)$ with $\theta(r)$, or equivalently with $\theta(\lambda)$, and we give a proof of it in the next section. The proof for the discrete formulation stated above is easily obtained following the same proof steps and it is left to the reader as an exercise. We give a sketch of the function θ for the two cases in Figure 3.1.

3.2.2 Full connectivity

We now ask for which choice of the parameters can we obtain a random grid where *all* vertices are connected a.a.s. Note this is not very meaningful for the site percolation model, as in this case each site is associated with a vertex that can be disconnected with probability $(1-p)$, so the network is connected if and only if all sites are occupied with probability one. A more interesting situation arises for edge percolation. Clearly, in this case p_n must tend to one as $n \to \infty$, and we are interested in discovering the exact rate required for convergence to a fully connected network.

The first theorem shows the correct scaling of p_n required for the number of isolated vertices to converge to a Poisson distribution.

Theorem 3.2.3 *Let W_n be the number of isolated vertices in G_n. Then W_n converges in distribution to a Poisson random variable with parameter $\lambda > 0$ if and only if*

$$n^2(1-p_n)^4 \to \lambda, \tag{3.10}$$

as $n \to \infty$.

Proof We start with the 'if' part. The main idea behind the proof is to use the Chen–Stein upper bound (3.5) on the total variation distance between the distribution of the number of isolated vertices in G_n and the Poisson distribution of parameter λ. First, to ensure we can apply such bound we need to show that isolated vertices are functions of independent a random variables. This follows from the *dual graph construction*, which

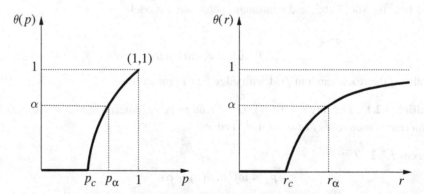

Fig. 3.1   Sketch of the discrete and continuous (boolean model) percolation function. This function is asymptotically equal to the fraction of connected nodes in G_n.

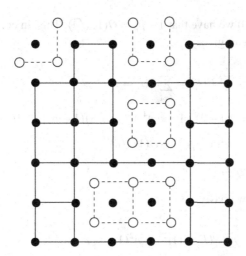

Fig. 3.2   Some configurations of isolated nodes in the random grid G_n. Dual paths are indicated with a dashed line.

is often used in percolation theory, and which we also used in Chapter 2. We refer to Figure 3.2. Let the dual graph of the $n \times n$ grid G_n be obtained by placing a vertex in each square of the grid (and also along the boundary) and joining two such vertices by an edge whenever the corresponding squares share a side. We draw an edge of the dual if it does not cross an edge of the original random grid, and delete it otherwise. It should be clear now that a node is isolated if and only if it is surrounded by a closed circuit in the dual graph, and hence node isolation events are increasing functions of independent random variables corresponding to the edges of the dual graph.

We can then proceed applying the Chen–Stein bound. Note that since isolated vertices are rare events, and most of them are independent, as $n \to \infty$, we expect this bound to tend to zero, which immediately implies convergence in distribution. Next we spell out the details; similar computations are necessary later, and it pays to see the details once.

Let I_i be the indicator random variable of node i being isolated, for $i = 1, \ldots, n^2$, so that

$$W_n = \sum_{i=1}^{n^2} I_i. \tag{3.11}$$

Let ∂G_n be the vertices on the boundary of G_n. We denote the four corner vertices by $\angle G_n$; the boundary vertices excluding corners by $\|G_n \equiv \partial G_n \setminus \angle G_n$; and the interior vertices by $\square G_n \equiv G_n \setminus \partial G_n$. We denote the expectation $E(W_n)$ by λ_n. We want to bound $d_{TV}(W_n, Po(\lambda))$, and we do this via an intermediate bound on $d_{TV}(W_n, Po(\lambda_n))$.

We start by computing some probabilities. For this computation, examining the corresponding dual lattice configurations depicted in Figure 3.2 might be helpful.

$$P(I_i = 1) = (1 - p_n)^4, \quad \text{for } i \in \square G_n,$$
$$P(I_i = 1) = (1 - p_n)^3, \quad \text{for } i \in \|G_n,$$
$$P(I_i = 1) = (1 - p_n)^2, \quad \text{for } i \in \angle G_n. \tag{3.12}$$

Note now that by (3.10) we have that $1 - p_n = O(1/\sqrt{n})$. This, in conjunction with (3.12) and a counting argument, gives

$$
\begin{aligned}
E(W_n) &= \sum_{i=1}^{n^2} E(I_i) = \sum_{i=1}^{n^2} P(I_i = 1) \\
&= (n-2)^2(1-p_n)^4 + (4n-8)(1-p_n)^3 + 4(1-p_n)^2 \\
&= n^2(1-p_n)^4 + O(1/\sqrt{n}) \\
&\to \lambda,
\end{aligned}
\tag{3.13}
$$

as $n \to \infty$. Similarly, we have

$$
\begin{aligned}
\sum_{i=1}^{n^2} (P(I_i = 1))^2 &= n^2(1-p_n)^8 + O(1/\sqrt{n}) \\
&= \lambda(1-p_n)^4 + O(1/\sqrt{n}) \\
&\to 0,
\end{aligned}
\tag{3.14}
$$

as $n \to \infty$. Finally, we also need to compute

$$
\begin{aligned}
E(W_n^2) &= E\left(\sum_\alpha I_\alpha \sum_\beta I_\beta \right) \\
&= E\left(\sum_\alpha I_\alpha + \sum_{\alpha \not\sim \beta} I_\alpha I_\beta + \sum_{\alpha \sim \beta} I_\alpha I_\beta \right),
\end{aligned}
\tag{3.15}
$$

where we have indicated with $\alpha \sim \beta$ the indices corresponding to neighbouring vertices, and $\alpha \not\sim \beta$ the indices corresponding to vertices that are not neighbouring nor the same. We proceed by evaluating the three sums in (3.15).

$$
E \sum_\alpha I_\alpha = E(W_n) \to \lambda,
\tag{3.16}
$$

$$
\begin{aligned}
E \sum_{\alpha \sim \beta} I_\alpha I_\beta &= \left\{ O(n^2)(1-p_n)^7 + O(n)[(1-p_n)^6 + (1-p_n)^5] + 8(1-p_n)^4 \right\} \\
&= O(1/n) \to 0,
\end{aligned}
\tag{3.17}
$$

where the different possible configurations of adjacent isolated nodes are depicted in Figure 3.3. Finally, the third sum yields

$$
\begin{aligned}
E \sum_{\alpha \not\sim \beta} I_\alpha I_\beta &= 2\left[\binom{n^2 - 4n + 4}{2} + O(n^2) \right](1-p_n)^8 + O(1/n) \\
&= n^4(1-p_n)^8 + o(1/n) \to \lambda^2,
\end{aligned}
\tag{3.18}
$$

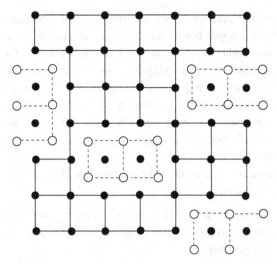

Fig. 3.3   Configurations of adjacent isolated nodes in the random grid G_n. Dual paths are indicated with a dashed line.

where the dominant term corresponds to the first order term of the configuration of all $\binom{n^4-4n+4}{2}$ isolated pairs not lying on the boundary, excluding the $O(n^2)$ isolated pairs that are adjacent to each other. Substituting (3.16), (3.17), (3.18) into (3.15), it follows that

$$\lim_{n\to\infty} \operatorname{Var} W_n = \lim_{n\to\infty} \left(E(W_n^2) - E(W_n)^2 \right) = \lambda. \tag{3.19}$$

By substituting (3.14) and (3.19) into (3.5), we finally obtain

$$\lim_{n\to\infty} d_{TV}(W_n, Po(\lambda_n)) = 0. \tag{3.20}$$

The proof is now completed by the observation that $d_{TV}(Po(\lambda), Po(\lambda_n))$ tends to zero as $n \to \infty$, since $\lambda_n \to \lambda$.

The 'only if' part of the theorem is easy. If $n^2(1-p_n)^4$ does not converge to λ, then the sequence either has a limit point $\lambda^* \neq \lambda$, or the sequence is unbounded. In the first case, by the first part of this proof, W_n converges along this subsequence in distribution to a Poisson random variable with parameter λ^*. In the second case, we have that W_n is (eventually) stochastically larger than any Poisson random variable with finite parameter. This means that for all λ there exists a large enough n such that $P(W_n \leq k) \leq P(Po(\lambda) \leq k)$, which clearly precludes convergence in distribution. □

The following result now follows without too much work; we ask for the details in the exercises.

Theorem 3.2.4   *Let $p_n = 1 - c_n/\sqrt{n}$ and let A_n be the event that there are no isolated nodes in G_n. We have that*

$$\lim_{n\to\infty} P(A_n) = e^{-c^4}, \tag{3.21}$$

if and only if $c_n \to c$ (where $c = \infty$ is allowed).

The careful reader will perhaps notice that the proof of Theorem 3.2.3 can also be used to obtain an explicit upper bound for the total variation distance between W_n and the Poisson distribution, rather than just showing that this distance tends to zero. When we replace c by a sequence c_n converging to c, then this explicit upper bound would prove Theorem 3.2.4 as well. In principle, if we had a rate of convergence of c_n to c, we could use the Chen–Stein upper bound to obtain a rate of convergence of the distributions as well. However, since the statements would become somewhat heavy, and since we do not want to assume any rate of convergence of c_n to c, we have opted for a simple approximation argument.

The next proposition articulates the relation between full connectivity and isolated nodes in the scaling of Theorem 3.2.4.

Proposition 3.2.5 *Suppose that $p_n = 1 - c_n/\sqrt{n}$, where $c_n \to c \in (0, \infty)$. Then w.h.p. G_n contains only isolated vertices, and in addition one component connecting together all vertices that are not isolated.*

Proof We use a counting argument in conjunction with the dual graph construction. First, we note that in order for the event described in the statement of the lemma not to occur, there must be either a self-avoiding path of length at least three in the dual graph starting at the boundary of the dual graph, or a self-avoiding path of length at least six starting in the interior of the dual graph; see Figure 3.3.

Let $P(\xi)$ be the probability of existence of a self-avoiding path of length at least ξ in the dual graph, starting from a given vertex. By the union bound, and since the number of paths of length k starting at a given vertex is bounded by $4 \cdot 3^{k-1}$, we have

$$P(\xi) \leq \sum_{k=\xi}^{\infty} 4 \cdot 3^{k-1} (1 - p_n)^k$$

$$= \frac{4}{3} \sum_{k=\xi}^{\infty} [3(1 - p_n)]^k$$

$$= \frac{4}{3} \sum_{k=\xi}^{\infty} \left(\frac{3c_n}{n^{\frac{1}{2}}} \right)^k$$

$$= \frac{4}{3} \frac{\left(\frac{3c_n}{n^{1/2}} \right)^\xi}{1 - \frac{3c_n}{n^{1/2}}}$$

$$= \frac{4}{3} (3c_n)^\xi \frac{n^{-\frac{1}{2}(\xi-1)}}{n^{\frac{1}{2}} - 3c_n}. \tag{3.22}$$

To rule out the possibility of a self-avoiding path of length at least three in the dual graph starting at the boundary of the dual graph, we are interested in $\xi = 3$, leading to an upper bound of $(4/3)(3c_n)^3 n^{-1}/(n^{1/2} - 3c_n)$. Since the number of boundary vertices is $4n - 4$, again applying the union bound it follows that the probability of a self-avoiding path of length three in the dual starting at the boundary, tends to zero as $n \to \infty$. To rule out the possibility of a self-avoiding path of length at least six in the dual graph starting in the interior of the dual graph, we take $\xi = 6$, leading to an upper bound of

$(4/3)(3c_n)^6 n^{-5/2}/(n^{1/2} - 3c_n)$. Since the number of such interior vertices is of the order n^2, the probability of a path of length at least 6 tends to zero as $n \to \infty$, completing the proof. $\qquad\square$

Combining Theorem 3.2.4 and Proposition 3.2.5, the following corollary follows. Again we ask for details in the exercises.

Corollary 3.2.6 *Let the edge probability $p_n = 1 - c_n/\sqrt{n}$. We have that G_n is connected w.h.p. if and only if $c_n \to 0$.*

Note that in the above corollary c_n is an arbitrary sequence that tends to zero. The corollary states that in order for the random grid to be fully connected, the edge probability must tend to one at a rate that scales slightly higher than the square root of the side length of the box. Here, 'slightly higher' is quantified by the rate of convergence to zero of the sequence c_n, which can be arbitrarily slow.

3.3 Boolean model

We now turn to the boolean model. Let X be a Poisson process of unit density on the plane. We consider the boolean random network model $(X, \lambda = 1, r > 0)$. As usual, we condition on a Poisson point being at the origin and let $\theta(r)$ be the probability that the origin is in an infinite connected component. We focus on the restriction $G_n(r)$ of the network formed by the vertices that are inside a $\sqrt{n} \times \sqrt{n}$ box $B_n \subset \mathbb{R}^2$. We call $N_\infty(B_n)$ the number of Poisson points in B_n that are part of an infinite connected component in the boolean model $(X, 1, r)$ over the whole plane. All the results we obtain also hold considering a box of unit length, density $\lambda = n$, and dividing all distance lengths by $\sqrt{n}$. We start by proving the following proposition.

Proposition 3.3.1 *We have $\theta(r) = E[N_\infty(B_1)]$.*

Proof Divide B_1 into m^2 subsquares s_i, $i = 1, \ldots, m^2$ of side length $1/m$ and define a random variable X_i^m that has value one if there is exactly one Poisson point in s_i that is also contained in an infinite connected component of the whole plane, and zero otherwise. Let $X_m = \sum_{i=1}^{m^2} X_i^m$. It should be clear that X_m is a non-decreasing sequence that tends to $N_\infty(B_1)$ as $m \to \infty$. Hence, by the monotone convergence theorem, we also have that

$$\lim_{m \to \infty} E(X_m) = E(N_\infty(B_1)). \qquad (3.23)$$

Let us now call s_i *full* if it contains exactly one Poisson point, and let A_i be the event that a Poisson point in s_i is part of an infinite component. Finally, call $\theta_m(r)$ the conditional probability $P(A_i | s_i \text{ full})$. It is not too hard to see that $\theta_m(r) \to \theta(r)$ as $m \to \infty$ (see the exercises). We have

$$E(X_i^m) = P(X_i^m = 1)$$
$$= P(A_i | s_i \text{ full}) P(s_i \text{ full})$$
$$= \theta_m(r) \left[\frac{1}{m^2} + o\left(\frac{1}{m^2}\right) \right]. \qquad (3.24)$$

It follows that

$$E(X_m) = m^2 E(X_i^m) = [1 + o(1)]\theta_m(r). \tag{3.25}$$

By taking the limit for $m \to \infty$ in (3.25), and using (3.23), we obtain

$$E[N_\infty(B_1)] = \lim_{m \to \infty} [1 + o(1)]\theta_m(r) = \theta(r). \tag{3.26}$$

$\square$

In a boolean model on the whole plane, the percolation function $\theta(r)$ represents the probability that a single point is in an infinite connected component. One might expect that the fraction of the points that are connected inside the box B_n is roughly equal to this function. This means that above criticality there is a value of the radius of the discs that allows a certain fraction of the nodes in B_n to be connected. On the other hand, if one wants to observe all nodes to be connected inside the box, then the radius of the discs must grow with the box size. We make these considerations precise below, starting with almost connectivity.

3.3.1 Almost connectivity

Definition 3.3.2 *For any $\alpha \in (0, 1)$, $G_n(r)$ is said to be α-almost connected if it contains a connected component of at least αn vertices.*

Note that in the above definition there are two differences from its discrete counterpart. First, we require αn vertices to be in a component rather than αn^2; this is simply because in this case the side length of the square is $\sqrt{n}$ rather than n. Second, the *average* number of vertices in B_n is n, while in the discrete case this number is exactly n. One could also state the definition requiring an α-fraction of the vertices in B_n to be connected. Indeed, this would be an equivalent formulation, since by the ergodic theorem (see Appendix A.3) the average number of Poisson points in B_n per unit area converges a.s. to its density as $n \to \infty$.

Theorem 3.3.3 *Let*

$$r_\alpha = \inf\{r \, ; \, \theta(r) > \alpha\}. \tag{3.27}$$

We have that for any $\alpha \in (0, 1)$, if $r > r_\alpha$, then $G_n(r)$ is α- almost connected a.a.s., while for $r < r_\alpha$ it is not.

Proof The proof is based on some geometric constructions. We start by showing that for $r > r_\alpha$, $G_n(r)$ is α-almost connected. We note that a sufficient condition to have a connected component in $G_n(r)$ containing αn vertices is the existence of a box $B_{\delta n}$ containing at least αn points of an infinite connected component, surrounded by a circuit of $G_n(r)$; see Figure 3.4. We will show that each of these events holds with arbitrarily high probability as $n \to \infty$. The union bound then immediately leads to the result.

Let us start by looking for a circuit of $G_n(r)$ surrounding $B_{\delta n}$. By Theorem 2.7.1 we have that if $r > r_c$, for any $0 < \delta < 1$, there exists a crossing path in a rectangle of sides $\sqrt{n} \times \sqrt{n}(1 - \sqrt{\delta})/2$, with high probability. We apply this result to the four rectangles surrounding $B_{\delta n}$, as depicted in Figure 3.5. We call CR_i, $i \in \{1, 2, 3, 4\}$ the four events

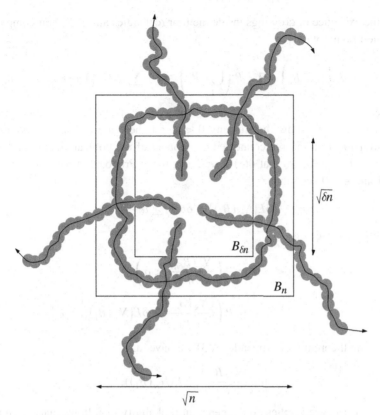

Fig. 3.4   Sufficient condition for almost connectivity.

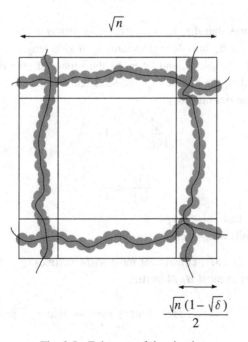

Fig. 3.5   Existence of the circuit.

denoting the existence of crossings inside the four rectangles and CR_i^c their complements. By the union bound we have

$$P\left(\bigcap_{i=1}^{4} CR_i\right) = 1 - P\left(\bigcup_{i=1}^{4} CR_i^c\right) \geq 1 - \sum_{i=1}^{4} P(CR_i^c) \longrightarrow 1, \qquad (3.28)$$

as $n \to \infty$.

The next step is to show that for any $0 < \alpha < 1$ there are at least αn points inside $B_{\delta n}$ that are part of an infinite connected component of the boolean model on the whole plane. We choose $r > r_\alpha$ so that $\theta(r) > \alpha$. Then, using Proposition 3.3.1, we can choose $0 < \delta < 1$ and $\epsilon > 0$ such that

$$\delta E[N_\infty(B_1)] = \delta\theta(r) \geq \alpha + \epsilon. \qquad (3.29)$$

From (3.29) it follows that

$$P(N_\infty(B_{\delta n}) < \alpha n) = P\left(\frac{N_\infty(B_{\delta n})}{n} < \alpha\right)$$

$$\leq P\left(\left|\frac{N_\infty(B_{\delta n})}{n} - \delta E[N_\infty(B_1)]\right| > \epsilon\right). \qquad (3.30)$$

By the ergodic theorem (see Appendix A.3) we have, a.s.,

$$\lim_{n \to \infty} \frac{N_\infty(B_{\delta n})}{\delta n} = E(N_\infty(B_1)). \qquad (3.31)$$

Since a.s. convergence implies convergence in probability, it follows that the right-hand side of (3.30) tends to zero as $n \to \infty$, which is what is needed to complete the first part of the proof.

We now need to show that if $r_c < r < r_\alpha$, then less than $\alpha = \theta(r_\alpha)$ nodes are connected. To do this, we partition B_n into M^2 subsquares s_i of side length $\sqrt{n}/M$ for some fixed $M > \sqrt{4/\alpha}$. Let $\delta \in (1 - \alpha/4, 1)$, and let w_i be the square of area $\delta|s_i| < \delta n\alpha/4$, placed at the centre of s_i, and A_i the annulus $s_i \setminus w_i$; see Figure 3.6.

Note that with these definitions,

$$\frac{|w_i|}{|s_i|} = \delta > 1 - \frac{\alpha}{4} \qquad (3.32)$$

and hence

$$\frac{|A_i|}{|s_i|} < \frac{\alpha}{4}. \qquad (3.33)$$

Finally, we also have that $|s_i| < n\alpha/4$.

We consider the following events.

(i)   Every s_i, $i = 1, \ldots, M^2$, contains at most $\alpha n/4$ vertices.
(ii) $\bigcup_{i=1}^{M^2} A_i$ contains at most $\alpha n/4$ vertices.
(iii) $N_\infty(B_n) < \alpha n$.
(iv) All annuli A_i, $i = 1, \ldots, M^2$, contain circuits that are part of the unbounded component.

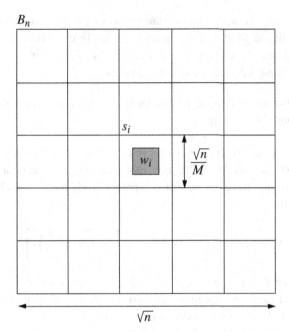

Fig. 3.6 Partition of the box and annuli construction.

Let us now look at the probabilities of these events. The ergodic theorem tells us that the number of points in a large region deviates from its mean by at most a small multiplicative factor. Hence, the event in (i) occurs w.h.p. since $|s_i| < n\alpha/4$. Since the union of the annuli A_i cover less than a fraction $\alpha/4$ of the square, the event in (ii) also occurs w.h.p., and similarly for the event in (iii). Event (iv) also occurs w.h.p. by the argument in the first part of the proof.

We claim now that the occurrence of events (i)–(iv) also implies that no component in B_n can have more than αn vertices. This is because each component that has vertices in two boxes w_i and w_j, $i \neq j$, also connects to the circuits in A_i and A_j that are in an infinite component. This implies by (iii) that it contains less than αn vertices. It remains to rule out the possibility of having components of size at least αn that are contained in $s_j \cup \bigcup_{i=1}^{M^2} A_i$, for some $j = 1, \ldots, M^2$. But by (i) and (ii) the number of vertices of this latter set is at most $2\alpha n/4 < \alpha n$ and this completes the proof. $\qquad \square$

3.3.2 Full connectivity

We now consider the situation where *all* points inside the box B_n form a connected cluster. We have previously seen that α-connectivity is achieved above the critical percolation radius r_α. The intuition in that case was that above criticality the infinite component invades the whole plane, including the area inside the box, and makes a fraction of the nodes in the box connected. This fraction is asymptotically equal to the value $\theta(r)$ of the percolation function. Now, if we want to observe a fully connected cluster inside a growing box, we clearly need to grow the radius of the discs with the box size. The problem is to identify at what rate this must be done. In the following, we see what is the exact threshold rate for asymptotic connectivity. We begin with a preliminary result that shows the required order of growth of the radius.

Theorem 3.3.4 *Let $\pi r_n^2 = \alpha \log n$. If $\alpha > 5\pi/4$, then $G_n(r)$ is connected w.h.p., while for $\alpha < 1/8$ it is not connected w.h.p.*

Proof We first show that $G_n(r)$ is not connected for $\alpha < 1/8$. Consider two concentric discs of radii r_n and $3r_n$ and let A^n be the event that there is at least one Poisson point inside the inner disc and there are no Poisson points inside the annulus between radii r_n and $3r_n$. We have that

$$P(A^n) = (1 - e^{-\pi r_n^2})e^{-8\pi r_n^2} = \left(\frac{1}{n}\right)^{8\alpha}\left[1 - \left(\frac{1}{n}\right)^{\alpha}\right], \qquad (3.34)$$

where we have used that $\pi r_n^2 = \alpha \log n$. Consider now 'packing' the box B_n with non-intersecting discs of radii $3r_n$. There are at least $\beta n/(\log n)$ of such discs that fit inside B_n, for some $\beta > 0$. A sufficient condition to avoid full connectivity of $G_n(r)$ is that A^n occurs inside at least one of these discs. Accordingly,

$$P(G_n(r) \text{ not connected}) \geq 1 - (1 - P(A^n))^{\frac{\beta n}{\log n}}. \qquad (3.35)$$

By (3.34) and exploiting the inequality $1 - p \leq e^{-p}$ that holds for any $p \in [0, 1]$, we have

$$(1 - P(A^n))^{\frac{\beta n}{\log n}} \leq \exp\left[-\frac{\beta n}{n^{8\alpha} \log n}\left(1 - \left(\frac{1}{n}\right)^{\alpha}\right)\right], \qquad (3.36)$$

which converges to zero for $\alpha < 1/8$. This completes the first part of the proof.

We now need to show that $G_n(r)$ is connected w.h.p. for $\alpha > 5\pi/4$. Let us partition B_n into subsquares S_i of area $\log n - \epsilon_n$, where $\epsilon_n > 0$ is chosen so that the partition is composed of an integer number $n/(\log n - \epsilon_n)$ of subsquares, and such that ϵ_n is the smallest such number. We call a subsquare *full* if it contains at least one Poisson point, and call it *empty* otherwise. The probability for a subsquare to be empty is $e^{-\log n + \epsilon_n}$, and we can compute the probability that every subsquare of B_n is full as

$$P\left(\bigcap_{i=1}^{\frac{n}{\log n - \epsilon_n}} S_i \text{ is full}\right) = \left(1 - e^{-\log n + \epsilon_n}\right)^{\frac{n}{\log n - \epsilon_n}}. \qquad (3.37)$$

Note that this latter probability tends to one as $n \to \infty$ since a little reflection shows that $\epsilon_n = o(1)$. We also note that any two points in adjacent subsquares are separated by at most a distance of $(5 \log n - 5\epsilon_n)^{1/2}$, which is the length of the diagonal of the rectangle formed by two adjacent subsquares. It follows that if

$$r_n > \frac{\sqrt{5 \log n - 5\epsilon_n}}{2}, \qquad (3.38)$$

then every point in a subsquare connects to all points in that subsquare and also to all points in all adjacent subsquares. This is the same condition as

$$\pi r_n^2 > \frac{\pi 5}{4}(5\log n - 5\epsilon_n). \qquad (3.39)$$

By dividing both sides of the inequality in (3.39) by $\log n$ and taking the limit for $n \to \infty$, it follows that for $\alpha > 5\pi/4$, points in adjacent subsquares are connected. Since by (3.37), w.h.p. every subsquare contains at least a Poisson point, the result follows. $\qquad \square$

The following theorem gives a stronger result regarding the precise rate of growth of the radius to obtain full connectivity.

Theorem 3.3.5 *Let $\pi(2r_n)^2 = \log n + \alpha_n$. Then $G_n(r_n)$ is connected w.h.p. if and only if $\alpha_n \to \infty$.*

Note the similarity with Corollary 3.2.6. The proof of this theorem is quite technical and rather long. We do not attempt to give all details here; these appear in the work of Penrose (1997). However, we want to highlight the main steps required to obtain a rigorous proof.

The first step is to show that isolated nodes do not arise w.h.p. inside the box if and only if $\alpha_n \to \infty$. This first step is shown by Proposition 3.3.6 and Proposition 3.3.7 below. The second step is to show that ruling out the possibility of having isolated nodes inside the box is equivalent to achieving full connectivity of all nodes inside the box. To do this, first we state in Theorem 3.3.8 that the longest edge of the nearest neighbour graph among the nodes in B_n has the same asymptotic behaviour as the longest edge of the tree connecting all nodes in B_n with minimum total edge length. Then, by Proposition 3.3.9, we show that this is also the same asymptotic behaviour of the critical radius for full connectivity of the boolean model inside the box.

The key to the first step is to approximate the sum of many low probability events, namely the events that a given node is isolated, by a Poisson distribution. One complication that arises in this case is given by boundary conditions. It is in principle possible that isolated nodes are low probability events close to the centre of the box, but that we can observe 'fake singletons' near the boundary of it. These are Poisson points that are connected on the infinite plane, but appear as singletons inside the box.

The key to the second step is a careful adaptation of the compression Theorem 2.5.6, valid on the whole plane, to a finite domain. The idea here is that at high density (or at large radii), if the cluster at the origin is finite, it is likely to be a singleton; then simply ruling out the possibility of observing isolated points inside a finite box should be sufficient to achieve full connectivity. However, even if we show that singletons cannot be observed anywhere in the box, and we know by the compression phenomenon that when radii are large no other isolated clusters can form, it is in principle possible to observe extremely large clusters that are not connected inside the box, but again only through paths outside the box. Theorem 2.5.6 simply does not forbid this possibility. Hence, the step from ruling out the presence of singletons inside the box to achieving full connectivity is not immediate. Finally, note that the compression theorem focuses only on the cluster at the origin, while we are interested in all points inside the box.

To adapt this theorem to a finite box and ensure that all we observe is likely to be a singleton, the probability of being a singleton conditioned on being in a component of constant size, must converge to one sufficiently fast when we consider the union of all points inside the box. All of these difficulties are carefully overcome in the work of Penrose (1997), and in the following we give an outline of this work. We first show that singletons asymptotically disappear, and then show the required steps to conclude that this is equivalent to having all finite clusters disappear.

Proposition 3.3.6   *If $\pi(2r_n)^2 = \log n + \alpha$, then the number of isolated nodes inside B_n converges in distribution to a Poisson random variable of parameter $\lambda = e^{-\alpha}$.*

We now state a slight variation of Proposition 3.3.6 that can be proven following the same arguments as in the proof of its discrete counterpart, Proposition 3.2.4. Note that this also shows that w.h.p. there are no isolated nodes inside B_n if and only if $\alpha_n \to \infty$.

Proposition 3.3.7   *Let $\pi(2r_n)^2 = \log n + \alpha_n$ and let A_n be the probability that there are no isolated nodes in B_n. We have that*

$$\lim_{n \to \infty} P(A_n) = e^{-e^{-\alpha}} \tag{3.40}$$

if and only if $\alpha_n \to \alpha$, where α can be infinity.

We now give a proof of Proposition 3.3.6 in the simpler case when B_n is a torus. This implies that we do not have special cases occurring near the boundary of the box, and that events inside B_n do not depend on the particular location inside the box.

Proof of Proposition 3.3.6 (Torus case)   The proof is based on a suitable discretisation of the space, followed by the evaluation of the limiting behaviour of the event that a node is isolated. Let us describe the discretisation first. Partition B_n into m^2 subsquares centred in $s_i \in \mathbb{R}^2$, $i = 1, \ldots, m^2$ of side length $\sqrt{n}/m$, and denote these subsquares by $V_i, i = 1, \ldots, m^2$. Let A_i^{mn} be the event that V_i contains exactly one Poisson point. For any fixed n, and any sequence $i_1, i_2, \ldots$, we have

$$\lim_{m \to \infty} \frac{P(A_{i_m}^{mn})}{n/m^2} = 1. \tag{3.41}$$

Note that for fixed m and n, the events A_i^{mn} are independent of each other, and that the limit above does not depend on the particular sequence (i_m).

We now turn to node isolation events. Let D_n be a disc of radius $2r_n$ such that $\pi(2r_n)^2 = \log n + \alpha$, centred at s_i. We call B_i^{mn} the event that the region of all subsquares intersecting $D_n \setminus V_i$ does not contain any Poisson point. For any fixed n, and any sequence $i_1, i_2, \ldots$, we have

$$\lim_{m \to \infty} \frac{P(B_{i_m}^{mn})}{e^{-\pi(2r_n)^2}} = 1. \tag{3.42}$$

Note that in (3.42) the limit does not depend on the particular sequence (i_m), because of the torus assumption. Note also that events B_i^{mn} are certainly independent of each other for boxes V_i centred at points s_i further than $5r_n$ apart, because in this case the corresponding discs D_n only intersect disjoint subsquares.

We define the following random variables for $i = 1, \ldots, m^2$:

$$
I_i^{mn} = \begin{cases} 1 & \text{if } A_i^{mn} \text{ and } B_i^{mn} \text{ occur,} \\ 0 & \text{otherwise,} \end{cases} \tag{3.43}
$$

$$
W_n^m = \sum_{i=1}^{m^2} I_i^{mn}, \quad W_n = \lim_{m \to \infty} W_n^m. \tag{3.44}
$$

Note that W_n indicates the number of isolated nodes in B_n. We now want to use the Chen–Stein bound in Theorem 3.1.5. Accordingly, we define a neighbourhood of dependence $\mathcal{N}_i$ for each $i \leq m^2$ as

$$
\mathcal{N}_i = \{j : |s_i - s_j| \leq 5r_n\}. \tag{3.45}
$$

Note that I_i^{mn} is independent of I_j^{mn} for all indices j outside the neighbourhood of independence of i. Writing I_i for I_i^{mn} and I_j for I_j^{mn}, we also define

$$
b_1 \equiv \sum_{i=1}^{m^2} \sum_{j \in \mathcal{N}_i} E(I_i)E(I_j),
$$

$$
b_2 \equiv \sum_{i=1}^{m^2} \sum_{j \in \mathcal{N}_i, j \neq i} E(I_i I_j). \tag{3.46}
$$

By Theorem 3.1.5 we have that

$$
d_{TV}(W_n^m, Po(\lambda)) \leq 2(b_1 + b_2), \tag{3.47}
$$

where $\lambda = E(W_n^m)$. Writing $a_m \sim_m b_m$ if $a_m/b_m \to 1$ as $m \to \infty$, using (3.41) and (3.42) we have

$$
\begin{aligned}
\lambda = E(W_n^m) &\sim_m n e^{-\pi(2r_n)^2} \\
&= e^{\log n - \pi(2r_n)^2} \\
&= e^{-\alpha}.
\end{aligned} \tag{3.48}
$$

Since the above result does not depend on n, we also have that

$$
\lim_{n \to \infty} \lim_{m \to \infty} E(W_n^m) = \lim_{n \to \infty} e^{-\alpha} = e^{-\alpha}. \tag{3.49}
$$

We now compute the right-hand side of (3.47). From (3.41) and (3.42) we have that

$$
E(I_i) \sim_m \frac{n}{m^2} e^{-\pi(2r_n)^2}. \tag{3.50}
$$

From this it follows that

$$
\begin{aligned}
\lim_{m \to \infty} b_1 &= \lim_{m \to \infty} \sum_{i=1}^{m^2} \left(\frac{n}{m^2} e^{-\pi(2r_n)^2} \right)^2 \frac{\pi(5r_n)^2}{n} m^2 \\
&= e^{-2\alpha} \frac{\pi(5r_n)^2}{n},
\end{aligned} \tag{3.51}
$$

which tends to 0 as $n \to \infty$.

We want to show similar behaviour for b_2. We start by noticing that $E(I_iI_j)$ is zero if two discs of radius $2r_n$, centred at s_i and s_j, cover each other's centres, because in this case the event A_i^{mn} cannot occur simultaneously with B_j^{mn}. Hence, we have

$$E(I_iI_j) = \begin{cases} 0 & \text{if } 2r_n > |s_i - s_j| \\ P(I_i = 1, I_j = 1) & \text{if } 2r_n < |s_i - s_j|. \end{cases} \tag{3.52}$$

We now look at the second possibility in (3.52). Let $D(r_n, x)$ be the area of the union of two discs of radius $2r_n$ with centres a distance x apart. Since B_i^{mn} and B_j^{mn} describe a region without Poisson points that tends to $D(r_n, |s_i - s_j|)$ as $m \to \infty$, for $2r_n < |s_i - s_j|$ we can write

$$E(I_iI_j) \sim_m \left(\frac{n}{m^2}\right)^2 \exp[-D(r_n, |s_i - s_j|)]. \tag{3.53}$$

We define an annular neighbourhood $\mathcal{A}_i$ for each $i \le m^2$ as

$$\mathcal{A}_i = \{j : 2r_n \le |s_i - s_j| \le 5r_n\}. \tag{3.54}$$

Combining (3.46), (3.52), and (3.53) we have

$$\lim_{m \to \infty} b_2 = \lim_{m \to \infty} \sum_{i=1}^{m^2} \sum_{j \in \mathcal{A}_i, j \ne i} \left(\frac{n}{m^2}\right)^2 \exp(-D(r_n, |s_i - s_j|))$$

$$= \lim_{m \to \infty} m^2 \sum_{j \in \mathcal{A}_i, j \ne i} \left(\frac{n}{m^2}\right)^2 \exp(-D(r_n, |s_i - s_j|))$$

$$= n \int_{2r_n \le |x| \le 5r_n} \exp(-D(r_n, |x|)) dx$$

$$\le n\pi(5r_n)^2 \exp\left(-\frac{3}{2}\pi(2r_n)^2\right), \tag{3.55}$$

where the last equality follows from the definition of the Riemann integral and the inequality follows from the geometry depicted in Figure 3.7. To see that this last expression tends to 0 as $n \to \infty$, substitute $\pi(2r_n)^2 = \log n + \alpha$ twice.

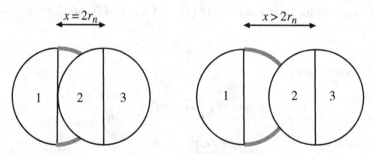

Fig. 3.7 The union of discs of radius $2r_n$ separated by a distance of at least $2r_n$ has area at least $3\pi(2r_n)^2/2$.

We have shown that both (3.51) and (3.55) tend to zero as $n \to \infty$, hence it follows from Theorem 3.1.5 that

$$\lim_{n \to \infty} \lim_{m \to \infty} d_{TV}(W_n^m, Po(\lambda)) = 0. \tag{3.56}$$

Since by definition W_n^m converges a.s. to W_n as $m \to \infty$, (3.48) and (3.56) imply that W_n converges in distribution to a Poisson random variable of parameter $e^{-\alpha}$ as $n \to \infty$. $\quad\square$

Having discussed the node-isolation results, we next need to relate these results to the question of full connectivity. In the discrete case we achieved this by a simple counting argument, but in a way we were just lucky there. In the current case, much more work is needed to formally establish this relation, and we no not give all details here, but we do sketch the approach now.

The two results above, given by Propositions 3.3.6 and 3.3.7, can be interpreted as the asymptotic almost sure behaviour of the length N_n of the *longest edge* of the nearest neighbour graph of the Poisson points inside B_n. Indeed, the transition from having one to having no isolated point when we let the radii grow clearly takes place when the point with the furthest nearest neighbour finally gets connected.

Now let the *Euclidean minimal spanning tree* (MST) of the Poisson points in B_n be the connected graph with these points as vertices and with minimum total edge length. Let M_n be the length of the longest edge of the MST. The following is a main result in Penrose (1997).

Theorem 3.3.8 *It is the case that*

$$\lim_{n \to \infty} P(M_n = N_n) = 1. \tag{3.57}$$

We also have the following geometric proposition.

Proposition 3.3.9 *If $r_n > M_n/2$, then $G_n(r_n)$ is connected; if $r_n < M_n/2$, then $G_n(r_n)$ is not connected.*

Proof Let $r_n > M_n/2$. Note that any two points connected by an edge in the MST are within distance $d \leq M_n$. It immediately follows that $MST \subseteq G_n(r_n)$ and hence $G_n(r_n)$ is connected. Now let $r_n < M_n/2$. By removing the longest edge (of length M_n) from the MST we obtain two disjoint vertex sets V_1 and V_2. Any edge joining these two sets must have length $d \geq M_n > 2r_n$, otherwise joining V_1 and V_2 would form a spanning tree shorter than MST, which is impossible. It follows that $G_n(r_n)$ cannot contain any edge joining V_1 and V_2 and it is therefore disconnected. $\quad\square$

Proof of Theorem 3.3.5 We combine the last two results. It follows from the last two results that w.h.p., if $r_n > N_n/2$, then the graph is connected, whereas for $r_n < N_n/2$ it is not. But we noted already that $r_n > N_n/2$ means that there are no isolated points, while $r_n < N_n/2$ implies that there are. This concludes the proof. $\quad\square$

3.4 Nearest neighbours; full connectivity

We now look at full connectivity of the nearest neighbour network $G_n(k)$ formed by Poisson points located inside B_n. As for the boolean model, we may expect that the number of connections per node needs to increase logarithmically with the side length of the box to reach full connectivity. We shall see that this is indeed the case. As usual, all results also hold considering a box of unit length, Poisson density n, and dividing all distance lengths by $\sqrt{n}$. We start by showing a preliminary lemma for Poisson processes that is an application of Stirling's formula (see Appendix A.2) and that will be useful in the proof of the main result. In the following, $|\cdot|$ denotes area.

Lemma 3.4.1 *Let $A_1(r), \ldots, A_N(r)$ be disjoint regions of the plane, and assume that each of their areas tends to infinity as $r \to \infty$. Let $\rho_1, \ldots, \rho_N \geq 0$ be such that $\rho_i > 0$ for some $i \leq N$, and $\rho_i|A_i(r)|$ are all integers. The probability that a Poisson process of density one on the plane has $\rho_i|A_i(r)|$ points in each region $A_i(r)$ is given by*

$$p = \exp\left(\sum_{i=1}^{N}(\rho_i - 1 - \rho_i \log \rho_i)|A_i(r)| + O(\log \sum_{i=1}^{N} \rho_i|A_i(r)|)\right), \quad (3.58)$$

as $r \to \infty$ and with the convention that $0 \log 0$ is zero.

Proof Let $n_i = \rho_i|A_i(r)|$. By independence, we have that

$$p = \prod_{i=1}^{N}\left(e^{-|A_i(r)|}\frac{|A_i(r)|^{n_i}}{n_i!}\right). \quad (3.59)$$

Taking logarithms of both sides, we can use Stirling's formula (see Appendix A.2) on the index set where the ρ_i are not zero, giving

$$\log p = \sum_{i:\rho_i \neq 0}\left(-|A_i(r)| + n_i \log |A_i(r)| - n_i \log n_i + n_i + O(\log n_i)\right)$$

$$- \sum_{i:\rho_i = 0}|A_i(r)|$$

$$= \sum_{i=1}^{N}(n_i - |A_i(r)| - n_i \log \rho_i) + O(\log \max n_i)$$

$$= \sum_{i=1}^{N}(\rho_i - 1 - \rho_i \log \rho_i)|A_i(r)| + O(\log \sum_{i=1}^{N} \rho_i|A_i(r)|). \quad (3.60)$$

$\square$

We now prove the main result of this section.

Theorem 3.4.2 *Let $k_n = \lfloor c \log n \rfloor$. If $c < 0.2739$, then $G_n(k_n)$ is not connected w.h.p.; if $c > 42.7$ then $G_n(k_n)$ is connected w.h.p.*

The bounds that we give for c in Theorem 3.4.2 can be strengthened considerably, at the expense of a more technical proof. Of course, this does not change the order of growth required for connectivity.

Proof of Theorem 3.4.2 First we show that $G_n(k_n)$ is not connected w.h.p. if $c < 0.2739$. The proof uses a discretisation of the space into regions of uniformly bounded size and then a geometric construction of an event occurring in such regions that ensures the network is disconnected. Finally it is shown that such an event occurs somewhere in the box B_n w.h.p., completing the proof.

Let us start with the discretisation. We describe this by fixing a radius r first, and in a second step we will choose $r = r_n$, and let the radius grow with the size of the box. Divide B_n into disjoint regions s_i of diameter $\epsilon_1 r \leq d_i \leq \epsilon_2 r$, for some $\epsilon_1, \epsilon_2 > 0$ and $r > 0$. These regions need not have the same shape. Consider three concentric discs D_1, D_3, and D_5, placed at the origin O and of radii r, $3r$, and $5r$ respectively. We call $A_1(r) \equiv D_1$, $A_2(r) \equiv (D_3 \setminus D_1)$, $\mathcal{A}(r) \equiv (D_5 \setminus D_3)$, and $A_i(r)$, $i = 3, \ldots, N$, the regions obtained by intersecting the annulus $\mathcal{A}(r)$ with all the regions s_i. See Figure 3.8 for a schematic picture of this construction. Note that since the area of $\mathcal{A}(r)$ is proportional to r^2 and the area of the s_i is bounded below by $(\epsilon_1 r)^2$, N is bounded above by some function of ϵ_1, uniformly in r.

We now describe a geometric construction that ensures the existence of disconnected clusters inside B_n. Let us assume that an integer number $\rho_i|A_i(r)|$ of Poisson points lie in each region $A_i(r)$, where $\rho_1 = 2\rho$, $\rho_2 = 0$ and $\rho_i = \rho$, $3 \leq i \leq N$, for some $\rho > 0$. It follows that the total number of Poisson points in the disc D_5 is $\sum \rho_i|A_i(r)| = 18\rho\pi r^2$. Note that this only makes sense if this number is an integer and we will later choose r and ρ such that this is the case. Given this configuration, consider a point x, placed at distance $r_x \geq 3r$ from O. Let D_x be the disc centred at x and of radius $r_x - (1 + \epsilon)r$, for some $\epsilon < \epsilon_2$ so small that

$$|D_x \cap \mathcal{A}(r)| \geq 2|A_1(r)|. \tag{3.61}$$

Note now that if one moves the point x radially outwards from the centre of $\mathcal{A}$, the discs D_x form a nested family. Hence (3.61) holds for all x. Note also that since the diameter of every cell s_i is at most $\epsilon_2 r < \epsilon r$, any $A_i(r)$ that intersects $D_x \cap \mathcal{A}(r)$ contains $\rho|A_i(r)|$

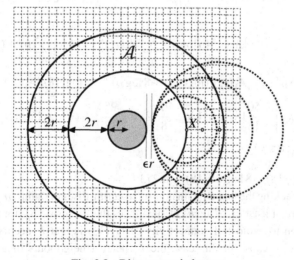

Fig. 3.8 Disconnected clusters.

points that are closer to x than any point of $A_1(r)$. From (3.61) it follows that any point $x \in \mathcal{A}(r)$ has at least $2\rho|A_1(r)|$ points in $D_x \cap \mathcal{A}(r)$ that are closer to itself than any point of $A_1(r)$. On the other hand, all points in $A_1(r)$ are closer to each other than to any point of $\mathcal{A}(r)$. Hence, letting $2k = \rho|A_1(r)| - 1$, the points in $A_1(r)$ form an isolated component.

We now evaluate the probability of observing the assumed geometric configuration described above. To do this we let the number of neighbours k and the radius r grow with n; it follows that also the size of the regions $A_i(r)$ grows to infinity. Accordingly, we let $k_n = \lfloor c \log n \rfloor$, $\rho = 25/18$, and $|A_1(r_n)| = \pi r_n^2 = (\lfloor c \log n \rfloor + 1)/2\rho = c \log n + o(\log n)$. Note that this choice is consistent with $2\rho|A_1(r)| = k + 1$ required in the geometric construction, and also that the desired number $18\rho\pi r^2$ of points inside D_5 is an integer and is chosen equal to the average number of Poisson points inside D_5, which is $25\pi r^2$. Let I_n be the event that each $A_i(r_n)$ contains exactly $\rho_i|A_i(r_n)|$ points of the Poisson process. By Lemma 3.4.1 we have that

$$P(I_n) = \exp\left[\sum_{i=1}^{N}\left(\rho_i|A_i(r_n)| - |A_i(r_n)| - |A_i(r_n)|\rho_i \log \rho_i\right)\right.$$
$$\left. + O\left(\log \sum_{i=1}^{N}\rho_i|A_i(r_n)|\right)\right], \tag{3.62}$$

and with some algebra we obtain

$$P(I_n) = \exp\left[2\rho|A_1(r_n)| + \rho\sum_{i=3}^{N}|A_i(r_n)| - 2\rho|A_1(r_n)|\log(2\rho)\right.$$
$$\left. - \rho\log\rho\sum_{i=3}^{N}|A_i(r_n)| - \sum_{i=1}^{N}|A_i(r_n)| + O\left(\log\sum_{i=1}^{N}\rho_i|A_i(r_n)|\right)\right]$$
$$= \exp\left[2\rho|A_1(r_n)| + 16\rho|A_1(r_n)| - 2\rho|A_1(r_n)|\log(2\rho)\right.$$
$$\left. - 16|A_1(r_n)|\rho\log\rho - \sum_{i=1}^{N}|A_i(r_n)| + O\left(\log\sum_{i=1}^{N}\rho_i|A_i(r_n)|\right)\right]$$
$$= \exp\left[-2\rho|A_1(r_n)|(\log(2\rho) + 8\log\rho) + O(\log(18\rho|A_1(r_n)|))\right]$$
$$= \exp\left[-\frac{50}{18}|A_1(r_n)|\left(\log\frac{50}{18} + 8\log\frac{25}{18}\right) + O(\log(25|A_1(r_n)|))\right]$$
$$= n^{-\frac{c}{c_0}+o(1)}, \tag{3.63}$$

where $c_0 = \log(50/18) + 8\log(25/18) \approx 0.2739$.

Consider now packing the box B_n with non-intersecting discs of radius $5r_n = 5[c_1 \log n + o(\log n)]$. There are at least $c_2 n/(\log n)$ of such discs that fit inside B_n, for some $c_2 > 0$. A sufficient condition to avoid full connectivity of $G_n(k_n)$ is that I_n occurs inside at least one of these discs, because in this case the k_n nearest neighbours of any point inside

$A_1(r_n)$ lie within $A_1(r_n)$, and the k nearest neighbours of any point inside $\mathcal{A}(r_n)$ lie outside $A_1(r_n) \cup A_2(r_n)$, and $A_2(r_n)$ does not contain any point. Accordingly,

$$P(G_n(k_n) \text{ not connected}) \geq 1 - (1 - P(I_n))^{\frac{c_2 n}{\log n}}. \tag{3.64}$$

By (3.63) and exploiting the inequality $1 - p \leq e^{-p}$ that holds for any $p \in [0, 1]$, we have

$$(1 - P(I_n))^{\frac{c_2 n}{\log n}} \leq \exp\left(-\frac{c_2 n}{n^{\frac{c}{c_0}} \log n}\right) \to 0, \tag{3.65}$$

for $c < c_0$, as $n \to \infty$, which completes the first part of the proof.

It remains now to be shown that $G_n(k_n)$ is connected w.h.p. for $k > \lfloor 42.7 \log n \rfloor$. We proceed in a similar fashion as in the proof of Theorem 3.3.4. Let us partition B_n into small subsquares s_i of area $\log n - \epsilon_n$, where $\epsilon_n > 0$ is chosen so that the partition is composed of an integer number $n/(\log n - \epsilon_n)$ of subsquares, and such that ϵ_n is minimal. We call a subsquare *full* if it contains at least a Poisson point, *empty* otherwise. The probability of a subsquare being empty is $e^{-\log n + \epsilon_n}$, and by (3.37) the event that all subsquares are full occurs w.h.p. We also note that any two points in adjacent subsquares are separated at most by a distance $(5 \log n - 5\epsilon_n)^{1/2}$, which is the diagonal of the rectangle formed by two adjacent subsquares. Let N_n be the number of Poisson points that lie in a disc of radius $\sqrt{5 \log n}$, and let $k = \lfloor 5\pi e \log n \rfloor < 42.7 \log n$. By Chernoff's bound (see Appendix A.4.3) we have

$$P(N_n > k) \leq e^{-5\pi \log n} = o(n^{-1}). \tag{3.66}$$

Consider now $n/(\log n - \epsilon_n)$ discs D_i of radius $\sqrt{5 \log n}$. Each D_i is centred at the lower left corner of s_i; we refer to Figure 3.9. Let A_n be the event that at least one of these discs contains more than k points. By (3.66) and the union bound we have

$$P(A_n) \leq o(n^{-1}) \frac{n}{\log n - \epsilon_n} \to 0. \tag{3.67}$$

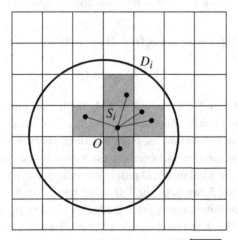

Fig. 3.9   There are at most k points inside the disc of radius $\sqrt{5 \log n}$ centred at O, and therefore a point inside the subsquare at the centre connects to all points in adjacent subsquares.

Now, since D_i contains all four subsquares adjacent to s_i, it follows from (3.67) that w.h.p. every point has at most k points within its adjacent subsquares. Hence, all points in adjacent subsquares are connected, and the proof is complete.   □

3.5 Critical node lifetimes

We end this chapter showing some scaling relations that are useful to determine critical node lifetimes in a random network. The main idea here is that nodes in the network have limited lifetime, and tend to become *inactive* over time. We wish to see how this new assumption can be incorporated in the scaling laws that we have derived in this chapter. Let us first give an informal picture: imagine we fix the system size n and let time t evolve. As nodes start progressively to fail, one might reasonably expect that there is a *critical time* t_n at which nodes with no active neighbours (i.e., *blind spots*) begin to appear in the network, and we are interested in finding the correct time scale at which this phenomenon can be observed.

To place the above picture into a rigorous framework, we proceed in two steps. First, we derive scaling laws for the number of blind spots in the network at a given time t. Then we fix n and let t evolve, and depending on the way we scale the radii, we obtain the time scale t_n at which the number of blind spots converges to a non-trivial distribution. This can effectively be viewed as the critical time at which blind spots can be first observed, and holds for a given failure probability distribution that is related to the battery drainage of the nodes. We first illustrate the situation for the discrete and then treat the continuum model.

Let us denote, as usual, by G_n the random $n \times n$ grid with edge probability p_n. For every n, all nodes have a common random lifetime distribution $q_n(t)$. That is, if we let T_i be the (random) failure time of node i, then for all $i \in G_n$, $P(T_i \leq t) = q_n(t)$. Node i is called *active* for $t < T_i$ and *inactive* for $t \geq T_i$. A blind spot is a node (either active or inactive) that is not connected to any active neighbours. For simplicity we consider a torus, but the results generalise to the square. Let $I_i(t)$ be the indicator random variable of the event that node i is a blind spot at time t. Now, $I_i(t) = 1$ if for all four neighbours of i, the neighbour is inactive or there is no connection. Accordingly, we have

$$P(I_i(t) = 1) = \left(q_n(t) + (1 - q_n(t))(1 - p_n) \right)^4$$

$$= (1 - p_n + p_n q_n(t))^4. \tag{3.68}$$

We can now apply the Chen–Stein method to study the limiting distribution as $n \to \infty$ of the sum of the dependent random variables $I_i(t)$, $i = 1, \ldots, n^2$, as we did in Theorem 3.2.3 for the distribution of the isolated nodes. Note that blind spots in this case are increasing functions of independent random variables corresponding to the edges of the dual graph and the state of the neighbours at time t. Omitting the tedious computations and proceeding exactly as in Theorem 3.2.3, we have the following result for the asymptotic behaviour of blind spots for a given failure rate $q_n(t)$ and edge probability p_n, which is the analogue of Theorems 3.2.3 and 3.2.4 in this dynamic setting. Considering a square instead of a torus, one needs to trivially modify (3.68) and patiently go through even more tedious, but essentially similar computations.

Theorem 3.5.1 *Let λ be a positive constant. The number of blind spots in G_n converges in distribution to a Poisson random variable of parameter λ if and only if*

$$n^2(1 - p_n + p_n q_n(t_n))^4 \to \lambda. \qquad (3.69)$$

Furthermore, letting A_n be the event that at time t_n there are no blind spots in G_n, if

$$p_n - p_n q_n(t_n) = 1 - \frac{c_n}{\sqrt{n}}, \qquad (3.70)$$

then

$$\lim_{n \to \infty} P(A_n) = e^{-\lambda} \qquad (3.71)$$

if and only if $c_n \to \lambda^{1/4}$.

We explicitly note that if $q_n(t) = 0$, then (3.70) reduces to the scaling of p_n given in Theorem 3.2.4.

We can now use Theorem 3.5.1 to derive the critical threshold time t_n at which blind spots begin to appear in the random network. To do this, we must fix n and let t evolve to infinity. Clearly, the critical time scale must depend on the given failure rate $q_n(t)$. Accordingly, we let

$$q_n(t) = 1 - e^{-t/\tau_n}, \qquad (3.72)$$

which captures the property that a given node is more likely to fail as time increases. In (3.72), τ_n can be interpreted as the time constant of the battery drainage of a given node in a network of size n^2. It is clear that other expressions of the failure rate, different from (3.72), can also be assumed. By substituting (3.72) into (3.70) we have that the critical time at which blind spots begin to appear in the random network is related to c_n as

$$\begin{aligned} t_n &= -\tau_n \log\left(\frac{1}{p_n} - \frac{c_n}{p_n\sqrt{n}}\right) \\ &= -\tau_n \log\left(\frac{1 - \frac{c_n}{\sqrt{n}}}{p_n}\right). \end{aligned} \qquad (3.73)$$

Some observations are now in order. Note that if $p_n = 1 - c_n/\sqrt{n}$, then $t_n = 0$, which is coherent with Theorem 3.2.4, stating that in this case there are isolated nodes even if all the nodes are active all the time, whenever $c_n \to c$. On the other hand, it is clear from (3.73) that if p_n approaches one at a faster rate than $1 - c_n/\sqrt{n}$, then the critical time scale required to observe blind spots increases. In practice, what happens is that a rate higher than what is required to avoid blind spots when all the nodes in the grid are active, provides some 'slack' that contrasts the effect of nodes actually becoming inactive over time. This effect of 'overwhelming connectivity' trading-off random node failures can also be appreciated in the continuum case, as we shall see next.

We assume that any Poisson point can become inactive before time t with probability $q_n(t)$. We also define $s_n(t) = 1 - q_n(t)$. The type of critical time – analogous to the discrete setting above – now strongly depends on the type of scaling that we use for the radii.

We first derive scaling laws for the number of *active* blind spots in the network.

Theorem 3.5.2 *If at the times t_n we have*

$$\pi(2r_n)^2 = \frac{\log\left(ns_n(t_n)\right) + \alpha_n}{s_n(t_n)}, \tag{3.74}$$

where $\alpha_n \to \alpha$ and $ns_n(t_n) \to \infty$, then the number of active blind spots in B_n at time t_n converges in distribution (as $n \to \infty$) to a Poisson random variable of parameter $e^{-\alpha}$.

Furthermore, if at the times t_n we have

$$\pi(2r_n)^2 = \frac{\log n + \alpha_n}{s_n(t_n)}, \tag{3.75}$$

where $\alpha_n \to \alpha$ and $s_n(t_n) \to s$, then the number of active blind spots in B_n at time t_n converges in distribution to a Poisson random variable of parameter $se^{-\alpha}$.

For blind spots that are not necessarily active, we have the following result.

Theorem 3.5.3 *If at the times t_n we have*

$$\pi(2r_n)^2 = \frac{\log n + \alpha_n}{s_n(t_n)}, \tag{3.76}$$

where $\alpha_n \to \alpha$ and

$$\frac{\log n}{\sqrt{n}s_n(t_n)} \to 0, \tag{3.77}$$

then the number of blind spots in B_n at time t_n converges in distribution to a Poisson random variable of parameter $\lambda = e^{-\alpha}$.

Some remarks are appropriate. Clearly, along any time sequence t_n one can choose r_n according to the above theorems to observe blind spot formation along the given time sequence. But one can also ask, given $s_n(t)$ and r_n, what is the corresponding critical time scale t_n for blind spot formation. Accordingly, in a similar fashion as in the discrete case, we can assume

$$s_n(t) = e^{-t/\tau_n}, \tag{3.78}$$

and substituting (3.78) into (3.75) or (3.76), we obtain the critical time for both kinds of blind spots to be

$$t_n = -\tau_n \log \frac{\log n + \alpha_n}{\pi(2r_n)^2}. \tag{3.79}$$

Notice that even if the critical times for the two kinds of blind spots are the same in the two cases, the corresponding Poisson parameters are different, one being $se^{-\alpha}$ and the other being $e^{-\alpha}$. In order to observe the same Poisson parameter, one needs to observe the system at different time scales, namely at the time scale provided by (3.79) in one case, and at the time scales provided (in implicit form) by (3.74). Furthermore, similar to the discrete case, the interpretation of (3.79) is again that if the denominator equals the numerator, the time at which we begin to observe blind spots is zero. In fact, in this case there is Poisson convergence even when there are no failures in the system, whenever $\alpha_n \to \alpha$. On the other hand, if the radius diverges at a faster rate than the critical threshold $\log n + \alpha_n$, then the ratio goes to zero and the critical time increases. Loosely speaking, the higher rate of divergence of the radius provides 'overwhelming connectivity', counteracting the effect of failures and increasing the time to observe blind spots.

Proof of Theorem 3.5.2 By Theorem 3.3.6 we have that for a continuum percolation model of unit density inside the box B_n, if $\pi(2r_n)^2 = \log n + \alpha_n$ with $\alpha_n \to \alpha$, then the number of isolated nodes converges to a Poisson distribution of parameter $e^{-\alpha}$. Writing s_n for $s_n(t_n)$, the same clearly holds for the continuum model in the box B_{ns_n} with density one, and a radius r_n satisfying $\pi(2r_n)^2 = \log(ns_n) + \alpha_n$, as long as $ns_n \to \infty$. Indeed, under this last condition this just constitutes a subsequence of our original sequence of models. When we now scale the latter sequence of models back to the original size B_n, that is, we multiply all lengths by $s_n^{-1/2}$, we obtain a sequence of models with density s_n and radii r_n given by

$$\pi(2r_n)^2 = \frac{\log(ns_n) + \alpha_n}{s_n}. \tag{3.80}$$

It immediately follows that in this model the number of isolated nodes converges to a Poisson distribution of parameter $e^{-\alpha}$, and the first claim follows.

To prove the second claim, we simply write

$$\frac{\log n + \alpha_n}{s_n} = \frac{\log(ns_n) + \alpha_n - \log s_n}{s_n}, \tag{3.81}$$

and then the claim follows from the previous result, with α_n replaced by $\alpha_n - \log s_n$. □

Proof of Theorem 3.5.3 We use the Chen–Stein bound (3.8) and the argument is a slight modification of the proof of Theorem 3.3.6. The required modification is to change the event of not having Poisson points in a region of radius $2r_n$, to the event of not having any *active* Poisson point in such a region and to go through the computations.

The proof is based on a suitable discretisation of the space, followed by the evaluation of the limiting behaviour of the event that a node is isolated. Let us describe the discretisation first. We start working on a torus and we partition B_n into m^2 subsquares (denoted by $V_i, i = 1, \ldots, m^2$) of side length $\sqrt{n}/m$, and centred in $c_1, \ldots, c_{m^2}$ respectively. Let A_i^{mn} be the event that V_i contains exactly one Poisson point. For any fixed n and any sequence $i_1, i_2, \ldots$, we have

$$\lim_{m \to \infty} \frac{P(A_{i_m}^{mn})}{\frac{n}{m^2}} = 1. \tag{3.82}$$

Note that, for fixed m, n, the events A_i^{mn} are independent of each other, and that the limit above does not depend on the particular sequence (i_m). We now turn to consider node isolation events, writing $s_n = s_n(t_n)$. Let D_n be a disc of radius $2r_n$ such that

$$\pi(2r_n)^2 = \frac{\log n + \alpha_n}{s_n}, \tag{3.83}$$

centred at c_i. We let B_i^{nm} be the event that the region of all subsquares intersecting $\{D_n \setminus V_i\}$ does not contain any active Poisson point. For any fixed n, and any sequence $i_1, i_2, \ldots$, we have

$$\lim_{m \to \infty} \frac{P(B_{i_m}^{nm})}{e^{-\pi(2r_n)^2 s_n}} = 1. \tag{3.84}$$

Note that in (3.84) the limit does not depend on the particular sequence (i_m), because of the torus assumption. Note also that events B_i^{nm} are certainly independent of each other for boxes V_i centred at points c_i further than $5r_n$ apart, because in this case the corresponding discs D_n only intersect disjoint subsquares.

We now define the following random variables

$$
I_i^{mn} = \begin{cases} 1 & \text{if } A_i^{mn} \text{ and } B_i^{nm} \text{ occur,} \\ 0 & \text{otherwise,} \end{cases} \tag{3.85}
$$

$$
W_n^m = \sum_{i=0}^{m^2} I_i^{nm}, \quad W_n = \lim_{m \to \infty} W_n^m. \tag{3.86}
$$

We want to use the Chen–Stein bound in Theorem 3.1.5. Accordingly, we define a neighbourhood of dependence $\mathcal{N}_i$ for each $i \leq m^2$ as

$$
\mathcal{N}_i = \{ j : |c_i - c_j| \leq 5r_n \}, \tag{3.87}
$$

Note that the I_i^{nm} is independent from I_j^{nm} for all indices outside the neighbourhood of dependence of i. Writing I_i for I_i^{nm} and I_j for I_j^{nm}, we also define

$$
b_1 \equiv \sum_{i=1}^{m^2} \sum_{j \in \mathcal{N}_i} E(I_i), E(I_j)
$$

$$
b_2 \equiv \sum_{i=1}^{m^2} \sum_{j \in \mathcal{N}_i, j \neq i} E(I_i I_j). \tag{3.88}
$$

By Theorem 3.1.5 we have that

$$
d_{TV}[W_n^m, Po(\lambda)] \leq 2(b_1 + b_2), \tag{3.89}
$$

where $\lambda \equiv E(W_n^m)$ can be computed as follows. Writing $a \sim_m b$ if $a/b \to 1$ as $m \to \infty$, using (3.82), (3.84) we have

$$
E(W_n^m) \sim_m ne^{-\pi(2r_n^2)s_n}
$$

$$
= e^{\log n - \frac{(\log n + \alpha_n)}{s_n} s_n}
$$

$$
= e^{-\alpha_n}. \tag{3.90}
$$

We then also have that

$$
\lim_{n \to \infty} \lim_{m \to \infty} E(W_n^m) = \lim_{n \to \infty} e^{-\alpha_n} = e^{-\alpha}. \tag{3.91}
$$

We now compute the right-hand side of (3.89). From (3.82) and (3.84) we have that

$$
E(I_i) \sim_m \frac{n}{m^2} e^{-\pi(2r_n)^2 s_n}. \tag{3.92}
$$

From which it follows that

$$\lim_{m \to \infty} b_1 = \lim_{m \to \infty} \sum_{i=1}^{m^2} \left(\frac{n}{m^2} e^{-\pi(2r_n)^2 s_n} \right)^2 \frac{\pi(5r_n)^2}{n} m^2$$

$$= e^{-2\alpha_n} \frac{\pi(5r_n)^2}{n}$$

$$= e^{-2\alpha_n} \frac{25}{4} \frac{\log n + \alpha_n}{ns_n}, \tag{3.93}$$

which by assumption tends to zero as $n \to \infty$.

For b_2, we start by noticing that $E(I_i I_j)$ is zero if two discs of radius $2r_n$, centred at c_i and c_j, cover each other's centres, because in this case the event A_i^{mn} cannot occur simultaneously with B_j^{mn}. Hence, we have

$$E(I_i I_j) = \begin{cases} 0 & \text{if } 2r_n > |c_i - c_j| \\ P(I_i = 1, I_j = 1) & \text{if } 2r_n < |c_i - c_j|. \end{cases} \tag{3.94}$$

We now look at the second term in (3.94). Let $D(r_n, x)$ be the area of the union of two discs of radius $2r_n$ with centres a distance x apart. Since B_i^{mn} and B_j^{mn} describe a region without active Poisson points, whose area tends to $D(r_n, |c_i - c_j|)$ as $m \to \infty$, for $2r_n < |c_i - c_j|$ we can write

$$E(I_i I_j) \sim_m \left(\frac{n}{m^2} \right)^2 \exp[-s_n D(r_n, |c_i - c_j|)]. \tag{3.95}$$

We define an annular neighbourhood $\mathcal{A}_i$ for each $i \leq m^2$ as

$$\mathcal{A}_i = \{j : 2r_n \leq |c_i - c_j| \leq 5r_n\}. \tag{3.96}$$

Combining (3.94) and (3.95), we have

$$\lim_{m \to \infty} b_2 = \lim_{m \to \infty} \sum_{i=1}^{m^2} \sum_{j \in \mathcal{A}_i, j \neq i} \left(\frac{n}{m^2} \right)^2 \exp[-s_n D(r_n, |c_i - c_j|)]$$

$$= \lim_{m \to \infty} m^2 \sum_{j \in \mathcal{A}_i, j \neq i} \left(\frac{n}{m^2} \right)^2 \exp[-s_n D(r_n, |c_i - c_j|)]$$

$$= n \int_{2r_n \leq |x| \leq 5r_n} \exp[-s_n D(r_n, |x|)] dx$$

$$\leq n\pi(5r_n)^2 \exp\left(-s_n \frac{3}{2} \pi(2r_n)^2 \right), \tag{3.97}$$

where the last equality follows from the definition of the Riemann integral. Substituting

$$\pi(2r_n)^2 = \frac{\log n + \alpha_n}{s_n}, \tag{3.98}$$

and using that $\log n/(\sqrt{n} s_n) \to 0$, we see that this tends to zero as $n \to \infty$. $\qquad \square$

3.6 A central limit theorem

We have seen throughout this chapter the Poisson distribution arising when we take the sum of a large number of mostly independent indicator random variables whose probability decays with n. We now want to mention another distribution that naturally arises when one sums many random variables whose probability does not decay, namely the Gaussian distribution.

We start by stating of the most basic version of the central limit theorem which can be found in any introductory probability textbook.

Theorem 3.6.1 *Let $X_1, X_2, \ldots$ be independent random variables with the same distribution, and suppose that their common expectation μ and variance σ^2 are both finite. Let $S_n = X_1 + \cdots + X_n$. We have that*

$$P\left(\frac{S_n - n\mu}{\sigma\sqrt{n}} \le x\right) \to P(N \le x), \qquad (3.99)$$

where N has a standard Gaussian distribution.

We are interested in a version of the central limit theorem arising for the number of isolated points in a random connection model. This should be contrasted with Poisson convergence of the number of isolated nodes discussed earlier in this chapter for the boolean model. We consider the sum of a large number of random variables, but their probability distributions do not change as the system size grows. Again, the main obstacle for these results to hold are dependencies arising in the model.

Let K be a bounded subset of the plane. Consider a sequence of positive real numbers λ_n with $\lambda_n/n^2 \to \lambda$, let X_n be a Poisson process on $\mathbb{R}^2$ with density λ_n and let g_n be the connection function defined by $g_n(x) = g(nx)$. Consider the sequence of Poisson random connection models (X_n, λ_n, g_n) on $\mathbb{R}^2$. Let $I_n(g)$ be the number of isolated vertices of (X_n, λ_n, g_n) in K. We then have the following result.

Theorem 3.6.2 *As $n \to \infty$, we have*

$$P\left(\frac{I_n(g) - E(I_n(g))}{\sqrt{Var(I_n(g))}} \le x\right) \to P(N \le x), \qquad (3.100)$$

where N has a standard Gaussian distribution.

3.7 Historical notes and further reading

The results on α-connectivity in the boolean model follow from Penrose and Pisztora (1996), who give general results for arbitrary dimension. The simpler two-dimensional version of Theorem 3.3.3 presented here follows the Master's thesis of van de Brug (2003). Theorem 3.3.5 is due to Penrose (1997). A less rigorous argument also appears in Gupta and Kumar (1998), missing some of the details we emphasised in Section 3.3.2. The argument we presented for full connectivity in nearest neighbour networks follows Balister, Bollobás, *et al.* (2005), who also provide tighter bounds on the constants in front of the logarithmic term than those we have shown here, as a well as a simple non-rigorous sketch showing the required logarithmic order of growth. Previously, bounds were given in Gonzáles-Barrios and Quiroz (2003), and Xue and Kumar (2004). Full connectivity of the random grid and critical node lifetimes appear in Franceschetti and Meester (2006a). A proof of Theorem 3.6.2 can be found in Meester and van de Brug (2004), who corrected a previous argument of Roy and Sarkar (2003).

Exercises

3.1 Check that the argument given in the proof of Proposition 3.2.5 does not go through for paths of length 4. Can you explain why?

3.2 Provide a proof for almost connectivity of the random grid model (Theorem 3.2.2).

3.3 Investigate asymptotic full connectivity on a rectangular grid $[0, 2n] \times [0, n]$, as $n \to \infty$.

3.4 In the proof of Theorem 3.3.4, we have stated in (3.37) that $\epsilon_n = o(1)$. Can you explain why this is so?

3.5 Explain why it is not interesting to consider almost connectivity in nearest neighbour networks.

3.6 Complete the proof of Proposition 3.2.4.

3.7 Provide a complete proof of Proposition 3.3.7.

3.8 Give a formal proof of the statement $\theta_m(r) \to \theta(r)$ in the proof of Proposition 3.3.1.

3.9 Give a full proof of Corollary 3.2.6.

4

More on phase transitions

In this chapter we examine the subcritical and the supercritical phase of a random network in more detail, with particular reference to bond percolation on the square lattice. The results presented lead to the exact determination of the critical probability of bond percolation on the square lattice, which equals $1/2$, and to the discovery of additional properties that are important building blocks for the study of information networks that are examined later in the book.

One peculiar feature of the supercritical phase is that in almost all models of interest there is only one giant cluster that spans the whole space. This almost immediately implies that any two points in space are connected with positive probability, uniformly bounded below. Furthermore, the infinite cluster quickly becomes extremely rich in disjoint paths, as p becomes strictly greater than p_c. So we can say, quite informally, that above criticality, there are many ways to percolate through the model. On the other hand, below criticality the cluster size distribution decays at least exponentially fast in all models of interest. This means that in this case, one can reach only up to a distance that is exponentially small.

To conclude the chapter we discuss an approximate form of phase transition that can be observed in networks of fixed size.

4.1 Preliminaries: Harris–FKG Inequality

We shall make frequent use of the Harris–FKG inequality, which is named after Harris (1960) and Fortuin, Kasteleyn and Ginibre (1971). This expresses positive correlations between increasing events.

Theorem 4.1.1 *If A, B are increasing events, then*

$$P_p(A \cap B) \geq P_p(A)P_p(B). \tag{4.1}$$

More generally, if X and Y are increasing random variables such that $E_p(X^2) < \infty$ and $E_p(Y^2) < \infty$, then

$$E_p(XY) \geq E_p(X)E_p(Y). \tag{4.2}$$

It is quite plausible that increasing events are positively correlated. For example, consider the event A_x that there exists a path between two points x_1 and x_2 on the random grid, and the event B_y that there exists a path between y_1 and y_2. If we know that A_x occurs, then it becomes more likely that B_y also occurs, as the path joining y_1 with y_2 can use some of the edges that are already there connecting x_1 and x_2. Despite this simple intuition, the FKG inequality is surprisingly hard to prove.

There are versions of the FKG inequality for continuous models as well. The reader can find proofs as well as general statements in Grimmett (1999) and Meester and Roy (1996).

4.2 Uniqueness of the infinite cluster

In the supercritical phase there is a.s. only one unbounded connected component. This remarkable result holds for edge and site percolation on the grid, boolean, nearest neighbours, and the random connection model, and in any dimension. However, it does not hold on the random tree: every vertex can be the root of an infinite component with positive probability, and an infinite component has infinitely many dead branches; it follows that there are a.s. infinitely many infinite components for $p > p_c$.

We give a proof of the uniqueness result for bond percolation on the square lattice. It is not difficult to see that a similar proof works for any dimension $d \geq 2$.

Theorem 4.2.1 *Let Q be the event that there exists at most one infinite connected component in the bond percolation model on the d-dimensional integer lattice. For all p we have $P_p(Q) = 1$.*

The proof is based on the consideration that it is impossible to embed a regular tree-like structure into the lattice in a stationary way, the point being that there are not enough vertices to accommodate such a tree. We exploit this idea using the fact that the size of the boundary of a box is of smaller order than the volume of the box. Thus, the proof can be adapted to different graph structures, however, there are certain graphs (for example a tree) where this approach does not work, as the boundary of a box centred at the origin is of the same order as the volume of the box.

We first prove two preliminary results, the first being an immediate consequence of ergodicity.

Lemma 4.2.2 *For all $0 < p < 1$, the number of infinite clusters on the random grid is an a.s. constant (which can also be infinity).*

Proof For all $0 \leq N \leq \infty$, let A_N be the event that the number of infinite clusters is equal to N. Notice that such an event is translation invariant. It follows by ergodicity that E_N has probability either zero or one. Therefore, for all p there must be a unique N such that $P_p(A_N) = 1$. □

Lemma 4.2.3 *For all $0 < p < 1$, the number of infinite clusters is a.s. either zero, one, or ∞.*

Proof Suppose the number of infinite clusters (which is an a.s. constant according to Lemma 4.2.2) equals k, where $2 \leq k < \infty$. Then there exists a (non-random) number m such that the box B_m is intersected by all these k clusters with positive probability. More precisely, let A be the event that the k infinite clusters all touch the boundary of B_m. For m large enough we have that $P_p(A) > 0$, and A depends only on the state of the bonds outside B_m. Let B be the event that all bonds inside B_m are present in the random grid.

It follows that $P_p(A \cap B) = P_p(A)P_p(B) > 0$ by independence. But if the event $A \cap B$ occurs, then there is only one infinite cluster, a contradiction since we assumed there were k of them with probability one. $\qquad\square$

It should be noticed, and it is given as an exercise, that the proof of Lemma 4.2.3 does not lead to a contradiction if one assumes the existence of infinitely many infinite clusters. We are now ready to give a proof of Theorem 4.2.1.

Proof of Theorem 4.2.1 Note that according to Lemma 4.2.3, and since $p > p_c$, we need only to rule out the possibility of having infinitely many infinite clusters.

We define $x \in \mathbb{Z}^2$ to be an *encounter point* if (i) x belongs to an infinite cluster $C(x)$, and (ii) the set $C(x)\setminus\{x\}$ has no finite components and exactly three infinite components (the 'branches' of x). Now suppose that there are infinitely many infinite clusters. We shall use a similar argument as in the proof of Lemma 4.2.3 to show that in this case the origin is an encounter point with probability $\epsilon > 0$, and so is any other vertex. This will then lead to a contradiction.

We refer to Figure 4.1. Denoting with $|\cdot|$ the L_1 distance, define D_m to be the 'diamond' centred at the origin and of radius m, that is, $D_m = \{x \in \mathbb{Z}^2 : |x| \le m\}$; and consider the event A_m that in the configuration outside D_m, there are at least three infinite clusters which intersect the boundary of D_m. Under our assumption, we clearly have

$$\lim_{m \to \infty} P_p(A_m) = 1. \qquad (4.3)$$

Hence it is possible to choose m so large that $P_p(A_m) > 0$, and we fix such a *non-random* value for m. Next, we consider the occurrence of a certain configuration inside D_m. We

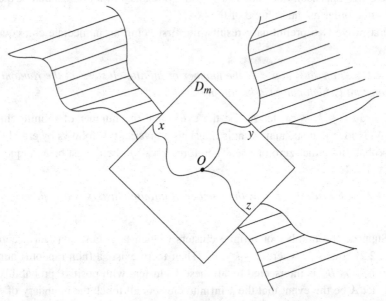

Fig. 4.1 Assuming there are at least three infinite clusters, the origin is an encounter point with positive probability.

start by noticing that there are three *non-random* points x, y, z on the boundary of D_m, such that with positive probability, x, y and z lie on three distinct infinite clusters outside D_m. Moreover, it is easy to see that it is always possible to connect any three points on the boundary of D_m to the origin using three non-intersecting paths inside the diamond; see Figure 4.1. We then let $J_{x,y,z}$ be the event that there exist such connecting paths for the points x, y, z and no other edges are present inside D_m. Then clearly we have $P_p(J_{x,y,z}) > 0$. Finally, using the independence of $J_{x,y,z}$ and A_m, we have

$$P_p(0 \text{ is an encounter point}) \geq P_p(J_{x,y,z})P_p(A_m)$$

$$= \epsilon, \tag{4.4}$$

where ϵ is some positive constant. By translation invariance, all points x have the same probability of being an encounter point, and we conclude that the expected number of encounter points in the box B_n is at least $n^2\epsilon$.

Now it is the case that if a box B_n contains k encounter points, then there will be at least $k+2$ vertices on the boundary of the box which belong to some branch of these encounter points. To see this, we refer to Figure 4.2. The three branches belonging to every encounter point are disjoint by definition, hence every encounter point is part of an infinite regular tree of degree three. There are at most k disjoint trees inside the box and let us order them in some arbitrary way. Let now r_i be the number of encounter points in the ith tree. It is easy to see that tree i intersects the boundary of the box in exactly $r_i + 2$ points, and since the total number of trees is at most k, the desired bound $k+2$ on the total number of intersections holds.

It immediately follows that the expected number of points on the boundary which are connected to an encounter point in B_n is at least $n^2\epsilon + 2$. This however is clearly impossible for large n, since the number of vertices on the boundary is only $4n$. $\quad\square$

An immediate consequence of Theorem 4.2.1 is the following.

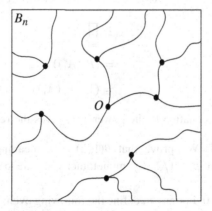

Fig. 4.2 Every encounter point inside the box is part of a tree of degree three.

Corollary 4.2.4   *For any $p > p_c$, any two points $x, y \in \mathbb{Z}^2$ are connected with probability*

$$P_p(x \leftrightarrow y) \geq \theta(p)^2. \tag{4.5}$$

Proof   There are only two ways for points x and y to be connected: they can either be both in the unique unbounded component C, or they can be part of the same finite component F. Hence, we can write

$$P_p(x \leftrightarrow y) \geq P_p(x, y \in C)$$
$$\geq \theta(p)^2, \tag{4.6}$$

where the last step follows from the FKG inequality.   □

Another interesting, but far less trivial, consequence of uniqueness is that the exact value of p_c on the square lattice is at least $1/2$. This improves the bound $p_c > 1/3$ that was given in Chapter 2 using a Peierls argument, and provides an important step towards the proof that $p_c = 1/2$.

Theorem 4.2.5   *For bond percolation on the two-dimensional square lattice, we have $p_c \geq 1/2$.*

In order to prove Theorem 4.2.5 we first show a preliminary technical lemma.

Lemma 4.2.6   (**Square-root trick**) *Let $A_1, A_2, \ldots, A_m$ be increasing events, all having the same probability. We have*

$$P_p(A_1) \geq 1 - \left(1 - P_p\left(\bigcup_{i=1}^{m} A_i\right)\right)^{1/m}. \tag{4.7}$$

Proof   By the FKG inequality, and letting A_i^c be the complement of event A_i, we have

$$1 - P_p\left(\bigcup_{i=1}^{m} A_i\right) = P_p\left(\bigcap_{i=1}^{m} A_i^c\right)$$
$$\geq \prod_{i=1}^{m} P_p(A_i^c)$$
$$= (P_p(A_1^c))^m$$
$$= (1 - P_p(A_1))^m. \tag{4.8}$$

Raising both sides of the equation to the power $1/m$ gives the result.   □

Proof of Theorem 4.2.5   We prove that $\theta(1/2) = 0$, meaning that a.s. there is no unbounded component at $p = 1/2$. By monotonicity of the percolation function, this implies that $p_c \geq 1/2$.

Assume that $\theta(1/2) > 0$. For any n, define the following events. Let $A^l(n)$ be the event that there exists an infinite path starting from some vertex on the left side of the box B_n, which uses no other vertex of B_n. Similarly, define $A^r(n)$, $A^t(n)$, $A^b(n)$ for the existence of analogous infinite paths starting from the right, top, and bottom sides of B_n and not using any other vertex of B_n beside the starting one. Notice that all these four events are increasing in p and that they have equal probability of occurrence. We call their union $U(n)$. Since we have assumed that $\theta(1/2) > 0$, we have that

$$\lim_{n \to \infty} P_{\frac{1}{2}}(U(n)) = 1. \tag{4.9}$$

By the square-root trick (Lemma 4.2.6) we have that each single event also occurs w.h.p. because

$$P_{\frac{1}{2}}(A^i(n)) \geq 1 - (1 - P_{\frac{1}{2}}(U(n)))^{\frac{1}{4}}, \quad \text{for } i = l, r, t, b, \tag{4.10}$$

which by (4.9) tends to one as $n \to \infty$. We can then choose N large enough such that

$$P_{\frac{1}{2}}(A^i(N)) \geq \frac{7}{8}, \quad \text{for } i = r, l, t, b. \tag{4.11}$$

We now shift our attention to the dual lattice. Let a dual box B_{nd} be defined as all the vertices of B_n shifted by $(1/2, 1/2)$; see Figure 4.3. We consider the events $A_d^i(n)$ that are the analogues of $A^i(n)$, but defined on the dual box. Since $p = 1/2$, these events have the same probability as before. We can then write

$$P_{\frac{1}{2}}(A_d^i(N)) = P_{\frac{1}{2}}(A^i(N)) \geq \frac{7}{8}, \quad \text{for } i = r, l, t, b. \tag{4.12}$$

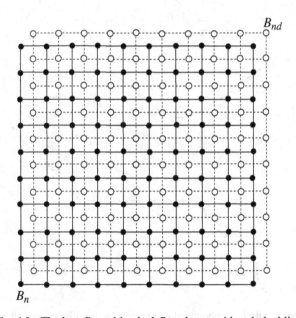

Fig. 4.3 The box B_n and its dual B_{nd}, drawn with a dashed line.

We now consider the event A that is a combination of two events occurring on the dual lattice and two on the original lattice. It is defined by

$$A = A^l(N) \cap A^r(N) \cap A_d^t(N) \cap A_d^b(N). \tag{4.13}$$

By the union bound and (4.12), we have that

$$P_{\frac{1}{2}}(A) = 1 - P_{\frac{1}{2}}(A^l(N)^c \cup A^r(N)^c \cup A_d^t(N)^c \cup A_d^b(N)^c)$$

$$\geq 1 - (P_{\frac{1}{2}}(A^l(N))^c + P_{\frac{1}{2}}(A^r(N))^c + P_{\frac{1}{2}}(A_d^t(N))^c + P_{\frac{1}{2}}(A_d^b(N))^c)$$

$$\geq \frac{1}{2}. \tag{4.14}$$

However, the geometry of the situation and Theorem 4.2.1 now lead to a contradiction, because they impose that $P_{\frac{1}{2}}(A) = 0$; see Figure 4.4. Event A implies that there are infinite paths starting from opposite sides of B_n, that do not use any other vertex of the box. However, any two points that lie on an infinite path must be connected, as they are part of the unique infinite component. But notice that connecting x_1 and x_2 creates a barrier between y_1 and y_2 that cannot be crossed, because otherwise there would be an intersection between an edge in the dual graph and one in the original graph, which

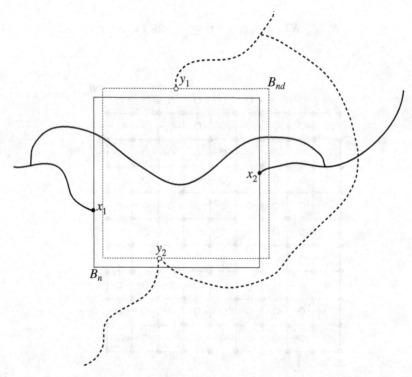

Fig. 4.4 Since there is only one unbounded component, x_1 must be connected to x_2. Similarly, in the dual graph, y_1 must be connected to y_2. This is a geometrically impossible situation.

is clearly impossible. We conclude that y_1 cannot be connected to y_2, which violates uniqueness of the infinite cluster in the dual lattice. □

4.3 Cluster size distribution and crossing paths

Beside the formation of an unbounded component, there are other properties that characterise the phase transition. Some of these relate to the probability of crossing a large box on the plane.

We start by considering the subcritical phase of the random grid. Let B_{2n} be a box of side length $2n$ centred at the origin, and let B_n denote a typical box of side length n. We denote by $0 \leftrightarrow \partial B_{2n}$ the event that there is a path connecting the origin to the boundary of B_{2n}, and with $B_n^{\leftrightarrow}$ the event that there is a crossing path connecting the left side of B_n with its right side. Our first bounds are easily obtained using a Peierls argument, which was also used in the proof of the phase transition in Chapter 2.

Proposition 4.3.1 *For $p < 1/3$ and for all n, we have*

$$P_p(0 \leftrightarrow \partial B_{2n}) \leq \frac{4}{3} e^{-\alpha(p)n}, \tag{4.15}$$

$$P_p(B_n^{\leftrightarrow}) \leq \frac{4}{3}(n+1) e^{-\alpha(p)n}, \tag{4.16}$$

where $\alpha(p) = -\log 3p$.

We explicitly note that since $p < 1/3$ both (4.15) and (4.16) tend to zero as $n \to \infty$.

Proof of Proposition 4.3.1 By (2.28) and since a path connecting the origin to the boundary of B_{2n} has length at least n, we have

$$P_p(0 \leftrightarrow \partial B_{2n}) \leq \frac{4}{3}(3p)^n = \frac{4}{3} e^{-\alpha(p)n}, \tag{4.17}$$

where $\alpha(p) = -\log 3p$.

We now prove the second part of the proposition. Let us order the vertices on the left side of the box B_n starting from the bottom, and let C_i be the event that there exists a crossing path starting from the ith vertex. There is a non-random index i_0 so that

$$P_p(C_{i_0}) \geq \frac{1}{n+1} P_p(B_n^{\leftrightarrow}). \tag{4.18}$$

Now choose the box B_n with this i_0th vertex being at the origin; see Figure 4.5 for an illustration of this construction with $i_0 = 3$. We then write

$$P_p(B_n^{\leftrightarrow}) \leq (n+1) P_p(C_{i_0})$$
$$\leq (n+1) P_p(0 \leftrightarrow \partial B_{2n})$$
$$\leq \frac{4}{3}(n+1) e^{-\alpha(p)n}. \tag{4.19}$$

□

Next, we see that a basic property of the square lattice leads to the following result.

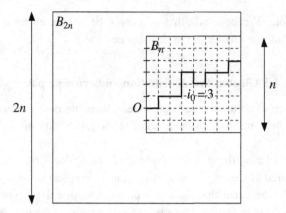

Fig. 4.5   The probability of crossing the box B_n starting from the third vertex from the bottom of B_n is less than the probability of reaching the boundary of box B_{2n} starting from its centre.

Proposition 4.3.2   *For $p > 2/3$ and for all n, we have*

$$P_p(B_n^{\leftrightarrow}) \geq 1 - \frac{4}{3}(n+1)e^{-\alpha(1-p)n}, \tag{4.20}$$

where $\alpha(\cdot)$ is as before.

Proof   Let us consider the box B_n, and the corresponding dual box S_n as depicted in Figure 4.6. Let $B_n^{\not\leftrightarrow}$ be the complement of the event that there is a left to right crossing path of B_n. This corresponds to the event that there exists a top to bottom crossing path

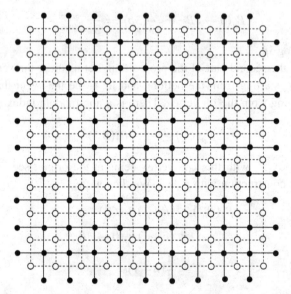

Fig. 4.6   The box B_n is drawn with a continuous line, the dual box S_n is drawn with a dashed line. Whenever there is not a top to bottom crossing in S_n, then there must be a left to right crossing in B_n.

in S_n. This last statement, which is immediate by inspection of Figure 4.6, can be given a complete topological proof; see Kesten (1982).

By rotating the box by 90° and applying Proposition 4.3.1 to the dual lattice, we have that, for all n,

$$P_p(B_n^{\not\leftrightarrow}) \leq (n+1)\frac{4}{3}e^{-\alpha(1-p)n}. \tag{4.21}$$

The result now follows immediately. $\qquad\qquad\qquad\qquad\qquad\qquad\qquad\qquad\square$

Perhaps not surprisingly, with much more work bounds based on weaker assumptions can be obtained.

Theorem 4.3.3 *For $p < p_c$ and for all n, there exists a $\beta(p) > 0$ such that $P_p(0 \leftrightarrow \partial B_{2n}) \leq e^{-\beta(p)n}$ and $P_p(B_n^{\leftrightarrow}) \leq (n+1)e^{-\beta(p)n}$.*

Theorem 4.3.4 *For $p > p_c$ and for all n, there exists a $\beta(1-p) > 0$ such that $P_p(B_n^{\leftrightarrow}) \geq 1 - (n+1)e^{-\beta(1-p)n}$.*

A corollary of Theorem 4.3.3 is that below criticality, the average number of vertices in the cluster at the origin is finite, see the exercises. Results are summarised in Figure 4.7, which depicts the transition behaviour. Below p_c the probability of reaching the boundary of a box of side $2n$ decays at least exponentially at rate $\beta(p)$, the probability of having a left to right crossing of B_n decays at least as fast as $(n+1)e^{-\beta(p)n}$, and a corresponding simple bound $\alpha(p)$ on the rate is found for $p < 1/3$. Similarly, above p_c the probability of having a left to right crossing of B_n converges at least as fast as $1 - (n+1)e^{-\beta(1-p)n}$, and a corresponding simple bound $\alpha(1-p)$ on the rate is found for $p > 2/3$. The careful reader might have noticed that for $p > p_c$ the probability of reaching the boundary of a box of side $2n$ does not go to one, but of course tends to $\theta(p)$.

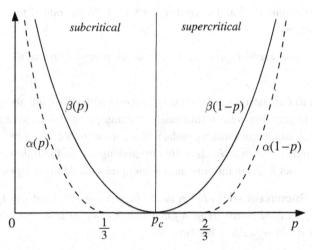

Fig. 4.7 In the subcritical region the probability of having a cluster of radius n decays at least exponentially with n at rate $\beta(p)$. In the supercritical region, the probability of having a crossing path in a box B_n increases at least as $1 - ne^{-\beta(1-p)n}$. The bounds $\alpha(p)$ and $\alpha(1-p)$ are easy to obtain using a Peierls argument.

The proof of Theorem 4.3.3 is quite long and we shall not give it here, we refer the reader to the book by Grimmett (1999). On the other hand, a proof of Theorem 4.3.4 immediately follows from Theorem 4.3.3, the dual lattice construction, and the observation that

$$p_{dual} = 1 - p < 1 - p_c \le p_c, \tag{4.22}$$

where the last inequality follows from Theorem 4.2.5.

We note that Theorem 4.3.3 has the following important corollary.

Corollary 4.3.5 *The critical probability for bond percolation on the square lattice is bounded above by $p_c \le 1/2$.*

Proof Assume that $p_c > 1/2$. This implies that at $p = 1/2$ the model is subcritical. Then, by Theorem 4.3.3 we have that $P_{1/2}(B_n^{\leftrightarrow})$ tends to zero as $n \to \infty$. A contradiction immediately arises by noticing that this probability is independent of n and equal to $1/2$. Perhaps this last statement merits some reflection. Notice that for $p = 1/2$ every realisation of the random network in B_n has the same probability, therefore to prove the claim it is enough to show that the number of outcomes in which there is a connection from left to right, is the same as the number of outcomes for which there is no such connection. Accordingly, we construct a one-to-one correspondence between these two possibilities. We recall that if the original network has a connection from left to right, then the dual network has no connection from top to bottom. If, on the other hand, the original network has no left–right connection, then there is a top to bottom connection in the dual. With this last observation in mind, the one-to-one correspondence becomes almost obvious: to each outcome of the original network which has a left to right connection, we associate the corresponding outcome in the dual and then rotate it by 90°. This gives the desired one-to-one correspondence, and finishes the proof.   □

Combining Theorem 4.2.5 and Corollary 4.3.5 we arrive at one of the most important results in percolation.

Theorem 4.3.6 *The critical probability for bond percolation on the square lattice equals $1/2$.*

We now turn to the question of how many crossing paths there are above criticality. It is reasonable to expect that we can find many crossing paths as we move away from p_c. As it turns out, the number of crossing paths will be proportional to n if we take a value of p only slightly higher than p_c. We show this by making use of the following general and useful lemma. The set $I_r(A)$ in this lemma is sometimes called the *interior of A of depth r*.

Lemma 4.3.7 **(Increment trick)** *Let A be an increasing event, and let $I_r(A)$ the event defined by the set of configurations in A for which A remains true even if we change the states of up to r arbitrary edges. We have*

$$1 - P_{p_2}(I_r(A)) \le \left(\frac{p_2}{p_2 - p_1}\right)^r (1 - P_{p_1}(A)), \tag{4.23}$$

for any $0 \le p_1 < p_2 \le 1$.

Proof We use the same coupling technique that we have introduced at the beginning of the book, see Theorem 2.2.2. We write

$$p_1 = p_2 \frac{p_1}{p_2},$$ (4.24)

where $p_1/p_2 < 1$. Let G_{p_2} be the random grid of edge probability p_2. By (4.24) we can obtain a realisation of G_{p_1} by deleting each edge independently from a realisation of G_{p_2}, with probability $(1 - p_1/p_2)$. On the other hand, it is also clear that the realisations of G_{p_1} and G_{p_2} are now coupled in the sense that G_{p_1} contains a subset of the edges of the original realisation of G_{p_2}.

We now note that if the event $I_r(A)$ does not hold for G_{p_2}, then there must be a set S of at most r edges in G_{p_2}, such that deleting all edges in S makes A false. The probability of a configuration where these edges are deleted when we construct a realisation of G_{p_1} starting from G_{p_2} is at least $(1 - p_1/p_2)^r$. Hence,

$$P(G_{p_1} \notin A \mid G_{p_2} \notin I_r(A)) \geq \left(\frac{p_2 - p_1}{p_2} \right)^r.$$ (4.25)

We then write

$$1 - P_{p_1}(A) = P_{p_1}(A^c)$$

$$\geq P(G_{p_1} \notin A, G_{p_2} \notin I_r(A))$$

$$= P(G_{p_1} \notin A \mid G_{p_2} \notin I_r(A)) \, P(G_{p_2} \notin I_r(A))$$

$$\geq \left(\frac{p_2 - p_1}{p_2} \right)^r P_{p_2}(I_r(A)^c)$$

$$= \left(\frac{p_2 - p_1}{p_2} \right)^r (1 - P_{p_2}(I_r(A))),$$ (4.26)

where the second inequality follows from (4.25). □

We can now prove the main result on the number of crossing paths in a box B_n above criticality.

Theorem 4.3.8 *Let M_n denote the maximal number of (pairwise) edge-disjoint left to right crossings of B_n. For any $p > p_c$ there exist positive constants $\delta = \delta(p)$ and $\gamma = \gamma(p)$ such that*

$$P_p(M_n \leq \delta n) \leq e^{-\gamma n}.$$ (4.27)

Proof Let A_n be the event of having a crossing from the left to the right side of B_n. Notice that this is an increasing event and a little thinking reveals that $I_r(A_n)$ is just the event that $r + 1$ edge-disjoint crossings exist. More formally, one can refer to the max-flow min-cut theorem to show this; see for example Wilson (1979).

For any $p > p_c$ we choose $p_c < p' < p$. By Lemma 4.3.7 and Theorem 4.3.4, there exists a $\beta(1 - p') > 0$, such that for any $\delta > 0$,

$$P_p(M_n \le \delta n) \le \left(\frac{p}{p - p'}\right)^{\delta n} e^{-\beta(1-p')n}. \tag{4.28}$$

We note that the above probability decays exponentially if

$$\gamma(p', p, \delta) = -\delta \log\left(\frac{p}{p - p'}\right) + \beta(1 - p') > 0. \tag{4.29}$$

We can now choose δ small enough to obtain the result. □

Theorem 4.3.8 says that when the system is above criticality there are a large number of crossing paths in the box B_n, namely, of the same order as the side length of the box.

We now want to discover more properties of these paths. It turns out that for all $\kappa > 0$, if we divide the box into rectangular slices of side length $n \times \kappa \log n$, then if we choose p sufficiently high, each of these rectangles contains at least a constant times $\log n$ number of disjoint crossings between the two shortest sides. This means that not only do there exist a large number of crossings of B_n, but also that these crossings behave almost as straight lines in connecting the two sides of the box, as they do not 'wiggle' more than a $\kappa \log n$ amount.

More formally, for any given $\kappa > 0$, let us partition B_n into rectangles R_n^i of sides $n \times (\kappa \log n - \epsilon_n)$, see Figure 4.8. We choose $\epsilon_n > 0$ as the smallest value such that the number of rectangles $n/(\kappa \log n - \epsilon_n)$ in the partition is an integer. It is easy to see that

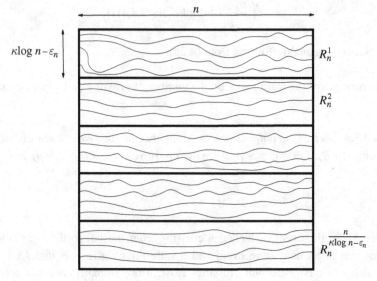

Fig. 4.8   There exist a large number of crossing paths in B_n that behave almost as straight lines.

$\epsilon_n = o(1)$ as $n \to \infty$. We let C_n^i be the maximal number of edge-disjoint left to right crossings of rectangle R_n^i and let $N_n = \min_i C_n^i$. The result is the following.

Theorem 4.3.9 *For all $\kappa > 0$ and $5/6 < p < 1$ satisfying $2 + \kappa \log(6(1-p)) < 0$, there exists a $\delta(\kappa, p) > 0$ such that*

$$\lim_{n \to \infty} P_p(N_n \le \delta \log n) = 0. \tag{4.30}$$

Proof Let $R_n^{i \leftrightarrow}$ be the event that there exists a left to right crossing of rectangle R_n^i. With reference to Figure 4.9, for all $p > 2/3$, and so in particular for $p > 5/6$, we have

$$P_p(R_n^{i \leftrightarrow}) \ge 1 - (n+1)P(0 \leftrightarrow \partial B_{2(\kappa \log n - \epsilon_n)})$$

$$\ge 1 - \frac{4}{3}(n+1)e^{-(\kappa \log n - \epsilon_n)(-\log(3(1-p)))}$$

$$= 1 - \frac{4}{3}(n+1)n^{\kappa \log(3(1-p))}(3(1-p))^{-\epsilon_n}, \tag{4.31}$$

where the first inequality follows from the same argument as in the proof of Proposition 4.3.2 and should be clear by looking at Figure 4.9, and the last inequality follows from Proposition 4.3.1 applied to the dual lattice. We now use the increment trick as we did in Theorem 4.3.8 to obtain $r = \delta \log n$ such crossings in R_n^m. For all $2/3 < p' < p < 1$, by Lemma 4.3.7 and (4.31), we have

$$P_p(C_n^i \le \delta \log n) \le \left(\frac{p}{p - p'}\right)^{\delta \log n} \frac{4}{3}(n+1)n^{\kappa \log(3(1-p'))}(3(1-p'))^{-\epsilon_n}. \tag{4.32}$$

Since $p > 5/6$, we have that letting $p' = 2p - 1 > 2/3$, so that

$$P_p(C_n^i \le \delta \log n) \le \frac{4}{3}(n+1)n^{\delta \log \frac{p}{1-p} + \kappa \log(6(1-p))}(6(1-p))^{-\epsilon_n}. \tag{4.33}$$

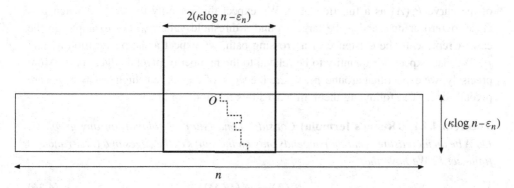

Fig. 4.9 Crossing the rectangle from left to right implies that there cannot be a top to bottom crossing in the dual graph. Hence, there cannot be a path from the centre to the lower side of any of the dual of the $n+1$ squares centred on the upper boundary of the rectangle.

We finally consider the probability of having less than $\delta \log n$ edge-disjoint left to right crossings in at least a rectangle R_n^i and show that this probability tends to zero. By the union bound and using (4.33), we have

$$P_p(N_n \le \delta \log n)$$

$$= P_p(\cup_i \{C_n^i \le \delta \log n\})$$

$$\le \frac{n}{\kappa \log n - \epsilon_n} \left(\frac{4}{3}(n+1) n^{\delta \log \frac{p}{1-p} + \kappa \log(6(1-p))} (6(1-p))^{-\epsilon_n} \right). \tag{4.34}$$

Since $\epsilon_n = o(1)$, we have that (4.34) tends to zero if

$$\delta \log \frac{p}{1-p} + \kappa \log(6(1-p)) + 2 < 0. \tag{4.35}$$

To complete the proof we can choose $\delta(\kappa, p)$ small enough so that (4.35) is satisfied. $\square$

4.4 Threshold behaviour of fixed size networks

We have seen how monotone events, such as the existence of crossing paths in a finite box, are intimately related to the phase transition phenomenon that can be observed on the whole plane. We have also seen in Chapter 3 how the phase transition phenomenon is an indication of the scaling behaviour of finite networks of increasing size.

We now ask what happens when we keep the system size fixed and let p change. Naturally, we expect also in this case to observe a behaviour that resembles the phase transition.

Here, we first present some indications of a similar threshold phenomenon arising for increasing events in the finite domain, and then we mention an approximate zero-one law that explains this phenomenon.

Let A be an increasing event. For such an event, define an edge e to be *pivotal* for A if the state of e determines the occurrence or non-occurrence of A. In other words, A occurs when e is present in the random network and does not occur otherwise, see Figure 4.10 for an example. Note that the pivotality of e does not depend on the state of e itself. We are interested in the rate of change of $P_p(A)$, that is, the derivative or slope of the curve $P_p(A)$ as a function of p. We expect this quantity to be small when p is close to zero or one, and to be large around a threshold value p_0. For example, in the case A represents the existence of a crossing path, we expect a sharp transition around p_c. We also expect this quantity to be related to the number of pivotal edges for A. More precisely, we expect that around p_c, where the slope of the curve is high, many edges are pivotal for A. The following theorem makes this relationship precise.

Theorem 4.4.1 (Russo's formula) *Consider bond (site) percolation on any graph G. Let A be an increasing event that depends only on the states of the edges in a (non-random) finite set $\mathcal{E}$. We have that*

$$\frac{d}{dp} P_p(A) = E_p(N(A)), \tag{4.36}$$

where $N(A)$ is the number of edges that are pivotal for A.

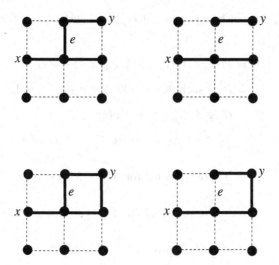

Fig. 4.10 Edge e is pivotal for $x \leftrightarrow y$ in the two top configurations, while it is not pivotal in the two bottom configurations.

Proof Let, as usual, G_p be the random network of edge probability p. We define the set $\{X(e) : e \in \mathcal{E}\}$ of i.i.d. uniform random variables in the set $[0, 1]$ and couple the edges of G_p with the outcome of $X(e)$ in the following way. We construct a realisation of G_p by drawing each edge e if $X(e) < p$, and delete it otherwise. Notice that in this way each edge is drawn independently with probability p.

We start by noticing the obvious fact

$$P_{p+\epsilon}(A) = P(\{G_{p+\epsilon} \in A \cap G_p \notin A\} \cup \{G_p \in A\})$$
$$= P(G_{p+\epsilon} \in A \cap G_p \notin A) + P_p(A). \qquad (4.37)$$

We also have

$$\frac{d}{dp} P_p(A) = \lim_{\epsilon \to 0^+} \frac{P_{p+\epsilon}(A) - P_p(A)}{\epsilon}$$
$$= \lim_{\epsilon \to 0^+} \frac{P(G_p \notin A, G_{p+\epsilon} \in A)}{\epsilon}, \qquad (4.38)$$

where the last equality follows from (4.37). We now note that if $G_{p+\epsilon} \in A$ and $G_p \notin A$ are both true, then $G_{p+\epsilon}$ must contain at least one edge (in $\mathcal{E}$) that is not in G_p. Let $E_{p\epsilon}$ be the number of such edges. Since $\mathcal{E}$ contains only finitely many edges, it is easy to see that

$$P(E_{p\epsilon} \geq 2) = o(\epsilon), \qquad (4.39)$$

as $\epsilon \to 0$. Since $E_{p\epsilon}$ cannot be equal to 0 if $G_{p+\epsilon} \in A$ and $G_p \notin A$ both occur, we may now write

$$
\begin{aligned}
P(G_p \notin A, G_{p+\epsilon} \in A) &= P(G_p \notin A, G_{p+\epsilon} \in A, E_{p\epsilon} = 1) + o(\epsilon) \\
&= P(\exists e \text{ such that } p \leq X(e) < p+\epsilon, E_{p\epsilon} = 1, \\
&\quad\quad G_p \notin A, G_{p+\epsilon} \in A) + o(\epsilon) \\
&= \sum_{e \in \mathcal{E}} P(p \leq X(e) < p+\epsilon, E_{p\epsilon} = 1, G_p \notin A, G_{p+\epsilon} \in A) + o(\epsilon) \\
&= \sum_{e \in \mathcal{E}} P(e \text{ is pivotal for } A, p \leq X(e) < p+\epsilon, E_{p\epsilon} = 1) + o(\epsilon) \\
&= \sum_{e \in \mathcal{E}} P(e \text{ is pivotal for } A, p \leq X(e) < p+\epsilon) + o(\epsilon) \\
&= \epsilon \sum_{e \in \mathcal{E}} P(e \text{ is pivotal for } A) + o(\epsilon), \quad\quad\quad (4.40)
\end{aligned}
$$

where the last equality follows from the independence of the state of an edge and it being pivotal or not. Dividing both sides by ϵ, taking the limit as $\epsilon \to 0^+$, and using (4.38) we have

$$
\begin{aligned}
\frac{d}{dp} P_p(A) &= \lim_{\epsilon \to 0^+} \left(\sum_{e \in \mathcal{E}} P(e \text{ is pivotal}) + o(1) \right) \\
&= E_p(N(A)). \quad\quad\quad\quad\quad\quad\quad\quad\quad (4.41)
\end{aligned}
$$

$\square$

We can get another indication of the behaviour of $P_p(A)$ from the following inequalities, which show that in the case the event A is the existence of a crossing path, the slope of $P_p(A)$ is zero for $p = 0$ and $p = 1$, and is at least one for $p = 1/2$.

Theorem 4.4.2 (Moore–Shannon inequalities) *Consider bond (site) percolation on any graph G. If A is an event that depends only on the edges in the set $\mathcal{E}$ of cardinality $m < \infty$, then*

$$
\frac{d}{dp} P_p(A) \geq \frac{P_p(A)(1 - P_p(A))}{p(1-p)}. \quad\quad\quad (4.42)
$$

Furthermore, if A is increasing, then

$$
\frac{d}{dp} P_p(A) \leq \sqrt{m \frac{P_p(A)(1 - P_p(A))}{p(1-p)}}. \quad\quad\quad (4.43)
$$

Proof We first claim that

$$
\frac{d}{dp} P_p(A) = \frac{1}{p(1-p)} \text{cov}_p(N, I_A), \quad\quad\quad (4.44)
$$

where I_A is the indicator variable of event A and N is the (random) number of edges of $\mathcal{E}$ that are good.

To see this, we write ω for a configuration in $\mathcal{E}$ and denote the outcome of a random variable X, when the configuration is ω by $X(\omega)$. Since

$$P_p(A) = \sum_\omega I_A(\omega) p^{N(\omega)} (1-p)^{m-N(\omega)} \qquad (4.45)$$

we can write

$$\frac{d}{dp} P_p(A) = \sum_\omega I_A(\omega) p^{N(\omega)} (1-p)^{m-N(\omega)} \left(\frac{N(\omega)}{p} - \frac{m-N(\omega)}{1-p} \right)$$

$$= \frac{1}{p(1-p)} \sum_\omega I_A(\omega) p^{N(\omega)} (1-p)^{m-N(\omega)} (N(\omega) - mp)$$

$$= \frac{1}{p(1-p)} \mathrm{cov}_p(N, I_A). \qquad (4.46)$$

Now apply the Cauchy–Schwartz inequality (see Appendix A.5) to the right side to obtain

$$\frac{d}{dp} P_p(A) \le \frac{1}{p(1-p)} \sqrt{\mathrm{var}_p(N) \mathrm{var}_p(I_A)}. \qquad (4.47)$$

Since N has a binomial distribution with parameters m and p we have $\mathrm{var}_p(N) = mp(1-p)$, and it is even easier to see that $\mathrm{var}_p(I_A) = P_p(A)(1 - P_p(A))$. Substituting this gives (4.42).

To prove (4.43), we apply the FKG inequality to find

$$\mathrm{cov}_p(N, I_A) = \mathrm{cov}_p(I_A, I_A) + \mathrm{cov}_p(N - I_A, I_A)$$

$$\ge \mathrm{var}_p(I_A), \qquad (4.48)$$

since both I_A and $N - I_A$ are increasing random variables. Substituting this in (4.44) finishes the proof. □

All of what we have learned so far about the function $P_p(A)$ indicates that finite networks have an 'S-shape' threshold behaviour that resembles a phase transition: for $p = 0$ and $p = 1$ the slope of the curve is zero, and for any p the slope is given by the expected number of pivotal edges. For crossing paths, we expect this number to be large around the critical percolation value p_c. Furthermore, we know that the slope at $p = 1/2$ is at least one.

We also know that a proper phase transition phenomenon can only be observed on the infinite plane. In Chapter 2 we have seen how this is a consequence of the Kolmogorov zero-one law, which states that tail events, i.e., events that are not affected by the outcome of any finite collection of independent random variables, have probability either zero or one.

The following result, Russo's approximate zero-one law, somehow connects Kolmogorov's zero-one law with the behaviour of finite systems. Stated informally, Russo's approximate zero-one law says that events that depend on the behaviour of a large but finite number of independent random variables, but are little influenced by the behaviour of each single random variable are almost always predictable.

A version of this law that holds for any graph G where the edges are marked independently with probability p is as follows.

Theorem 4.4.3 (**Russo's approximate zero-one law**) *Consider bond (site) percolation on any graph G. Let A be an increasing event. For any $\epsilon > 0$ there exists a $\delta > 0$ such that if for all e and for all p in $[0, 1]$, $P_p(e$ is pivotal$) < \delta$, then there exists a $0 < p_0 < 1$ such that*

$$\text{for all } p \leq p_0 - \epsilon \quad P_p(A) \leq \epsilon,$$

$$\text{for all } p \geq p_0 + \epsilon \quad P_p(A) \geq 1 - \epsilon. \tag{4.49}$$

Note that in the statement above the probability that one particular edge is pivotal must be sufficiently small for (4.49) to hold. One can expect that for crossing events in a box B_n this condition is satisfied when n is large enough and p is above the critical threshold p_c. Indeed, note that for tail events defined on the infinite plane, the probability of an edge to be pivotal is always zero.

A natural question to ask is for a finite network of size n, how large should n be to appreciate a sharp threshold phenomenon? This of course depends on the given δ required to satisfy the approximate zero-one law. It turns out that it is possible to directly relate δ to the width of the transition as follows.

Theorem 4.4.4 *Consider bond (site) percolation on any graph G_n of n vertices. Any increasing event A that depends only on the state of the edges (sites) of G_n has a sharp threshold, namely, if $P_p(A) > \epsilon$ then $P_q(A) > 1 - \epsilon$ for $q = p + \delta(\epsilon, n)$, where $\delta(\epsilon, n) = O\left(\log(1/2\epsilon)/(\log n)\right)$ as $n \to \infty$.*

We remark that Theorem 4.4.4 can also be extended to the boolean model network over a finite box B_n, where the radius (or the density) of the discs plays the role of the parameter p.

Finally, we want to make some comments on Figure 4.11. Let A be the event of having a crossing in the edge percolation model on the box B_n. Russo's law tells us that a transition occurs between $p_0 - \epsilon$ and $p_0 + \epsilon$. Moreover, since $P_{1/2}(A) = 1/2$, the Moore–Shannon inequalities tell us that the curve at $p = 1/2$ has at least a 45° angle.

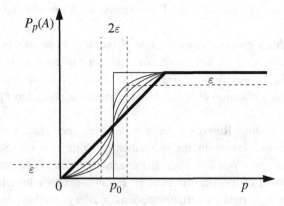

Fig. 4.11 The approximate zero-one law.

Russo's formula tells us that the slope equals the average number of pivotal edges for A, which we expect to attain a maximum at $p = 1/2$. Finally, Theorem 4.4.4 tells us that for any given ϵ, as n increases, the width of the transition tends to zero, which indicates that the curves in the figure tend to approximate more and more a step function at criticality, as the system size increases.

4.5 Historical notes and further reading

The Harris–FKG inequality appears in Harris (1960) and was later put in a more general context by Fortuin, Kasteleyn, and Ginibre (1971). Another useful correlation inequality that is applied in the opposite direction, requiring a disjoint set of edges, is the BK inequality which is named after van den Berg and Kesten (1985). Several refinements of these inequalities are available in the statistical physics literature under the general framework of correlation inequalities, see Grimmett (1999) and references therein. The uniqueness theorem 4.2.1 was first proven by Aizenman, Kesten, and Newman (1987). The simpler proof that we presented here is a wonderful advancement of Burton and Keane (1989), which, as we discussed, can be applied to different graph structures, and in any dimension. Uniqueness of the infinite cluster in the boolean model was proven by Meester and Roy (1994) and in the nearest neighbour networks by Häggström and Meester (1996).

In his classic paper, Harris (1960) proved that $p_c \geq 1/2$. The geometric argument we have given here used the uniqueness result and is due to Zhang (1988). The square-root trick appears in Cox and Durrett (1988). Only 20 years after Harris, Kesten (1980) was able to prove that $p_c \geq 1/2$, building on advancements by Russo (1978) and Seymour and Welsh (1978). Such advancements also contain the geometric ingredients on crossing paths that we have presented here. The increment trick in Lemma 4.3.7 was proven by Aizenman, Chayes, *et al.* (1983). Its application to obtain the result in Theorem 4.3.8 is also discussed in Grimmett (1999).

Russo's formula is named after Russo (1981). An earlier version appeared in the context of reliability theory, which is an applied mathematics field that intersects with percolation theory, pioneered by Moore and Shannon (1956) and Barlow and Proschan (1965). Russo's approximate zero-one law appears in Russo (1982), and was later generalised by Talagrand (1994). Theorem 4.4.4 follows from Friedgut and Kalai (1996). An extension to the boolean model when n points are uniformly distributed in a square is given by Goel, Rai, and Krishnamachari (2005).

Exercises

4.1 Let $|C|$ denote the number of vertices in the cluster at the origin. Prove that for $p > p_c$, $E(|C|) = \infty$, while for $p < p_c$, $E(|C|) < \infty$.

4.2 Provide an upper bound for $P(|C| > n)$.

4.3 Prove the first part of Theorem 4.3.4.

4.4   Prove the FKG inequality using induction on n and assuming that A and B depend only on finitely many edges. (Warning: this might be tricky).

4.5   Explain where the proof of Lemma 4.2.3 breaks down if one assumes $k = \infty$.

4.6   We have used ergodicity to prove Lemma 4.2.2. Explain why it is not possible to apply Kolmogorov's zero-one law to prove the result in this case.

4.7   Provide a complete proof that a tree of n vertices placed in a finite box, whose branches are connected to infinity, intersects the boundary of the box in exactly $n+2$ points. Notice that this property has been used in the proof of Theorem 4.2.1.

4.8   Explain why it is not immediately possible to replace the diamond D_m in the proof of Theorem 4.2.1 with a box B_m.

4.9   Explain why the uniqueness proof on Section 4.2 does not work on a tree. How many infinite components do you think there are on a tree, when $p > p_c$?

4.10 Consider bond percolation on the two-dimensional integer lattice, and take $p = 1/2$. Show that the probability that two given nearest neighbours are in the same connected component is equal to $3/4$.

5

Information flow in random networks

In the words of Hungarian mathematician Alfréd Rényi, 'the mathematical theory of information came into being when it was realised that the *flow of information* can be expressed numerically in the same way as distance, time, mass, temperature...'[1]

In this chapter, we are interested in the dynamics of the information flow in a random network. To make precise statements about this, we first need to introduce some information-theoretic concepts to clarify – from a mathematical perspective – the notion of information itself and that of communication rate. We shall see that the communication rate between pairs of nodes in the network depends on their (random) positions and on their transmission strategies. We consider two scenarios: in the first one, only two nodes wish to communicate and all the others help by relaying information; in the second case, different pairs of nodes wish to communicate simultaneously. We compute upper and lower bounds on achievable rates in the two cases, by exploiting some structural properties of random graphs that we have studied earlier. We take a statistical physics approach, in the sense that we derive scaling limits of achievable rates for large network sizes.

5.1 Information-theoretic preliminaries

The topics of this section only scratch the surface of what is a large field of study; we only discuss those topics that are needed for our purposes. The interested reader may consult specific information-theory textbooks, such as McEliece (2004), and Cover and Thomas (2006), for a more in-depth study.

The act of communication can be interpreted as altering the state of the receiver due to a corresponding action of the transmitter. This effect implies that some information has been transferred between the two. Let us be a little more formal and assume we have an index set $I = \{1, 2, \ldots, M\}$ of possible states we wish to communicate. First, we want to define the amount of information associated with one element of this set.

Definition 5.1.1 *The information of the set $I = \{1, 2, \ldots, M\}$ is the minimum number of successive binary choices needed to distinguish any one, among all elements of the set, by recursively splitting it into halves. This is given by* $\log M$ *bits, where the logarithm is in base 2 and rounded up to the next integer.*

---

[1] Quote from 'A diary on information theory'. John Wiley & Sons, Ltd, 1984.

Once we have an idea of the amount of information one element of the index set carries, we call the 'space' between two nodes in our network a *channel* and we can proceed by describing the act of communication over this channel. To do this, the M states must be put into a format suitable for transmission. We assume that a channel can transmit *symbols* taken from a set S, that represent some kind of physical quantity, electrical signal levels for example. Accordingly, the M states are encoded into codewords, each of length m symbols, using an encoding function

$$X^m : \{1, 2, \ldots, M\} \longrightarrow S^m, \tag{5.1}$$

yielding codewords $X^m(1), X^m(2), \ldots, X^m(M)$. Different codewords, representing different states of the set I, can then be transmitted over the channel. In this way each codeword, identifying one element of I, carries $\log M$ bits of information across the channel using m symbols, and we say that the *rate* of this (M, m) coding process for communication is

$$R = \frac{\log M}{m} \text{ bits per symbol.} \tag{5.2}$$

Notice that since the channel accepts $|S|$ possible symbols, where $|\cdot|$ denotes cardinality, $|S|^m$ is the total number of different words of length m that can be transmitted over the channel. In order to associate them with M distinct states we need $|S|^m \geq M$, from which it follows that

$$m \geq \frac{\log M}{\log |S|}. \tag{5.3}$$

Substituting (5.3) into (5.2), we obtain the following upper bound on the rate,

$$R \leq \log |S|. \tag{5.4}$$

In the special case when the channel can transmit binary digits as symbols, $|S| = 2$ and (5.4) reduces to $R \leq 1$, which simply states that one needs at least to transmit a bit in order to receive a bit of information.

5.1.1 Channel capacity

From what we have learned so far, it seems that given a set of symbols S of cardinality $|S|$ and a channel that can carry such symbols, we can transmit information at maximum rate $\log |S|$, by simply encoding each information state into one transmission symbol. It turns out, however, that this is not possible in practice, because the physical process of communication is subject to some constraints, which limit the code rate.

The first constraint rules out the possibility of having arbitrary inputs. The states we wish to communicate are modelled as random and induce a certain probability distribution on the codewords $X^m = (X_1, \ldots, X_m)$. This distribution is subject to the mean square constraint

$$E(X_i^2) \leq \beta, \text{ for all } i, \tag{5.5}$$

where β is a given constant. Let us accept this constraint for the moment as a modelling assumption. Later, we shall give it a physical explanation in terms of power available for transmission. The second constraint is given by the noise associated with the transmission

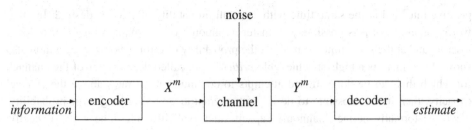

Fig. 5.1 The communication system.

process. Because of the noise, only a 'corrupt' version of the symbols can be received, and a decoding function $g : S^m \longrightarrow \{1, 2, \ldots, M\}$ can only assign a 'guess' to each received word. A schematic representation of this process is given in Figure 5.1.

Let us now see how the two constraints above limit the code rate. We first give an informal picture. If one encodes each information state into one transmission symbol, then because of the noise distinct symbols (e.g. distinct electrical signal levels) must take sufficiently different values to be distinguishable. It follows that the symbol values must be spread over a large interval, and (5.5) does not allow this.

One strategy to combat the noise while keeping the symbol values close to each other, could be to use longer codewords to describe elements of the same information set. By repeating each symbol multiple times, or by adding some extra symbols to the word in some intelligent way, one can expect that the added redundancy would guarantee a smaller probability of error in the decoding process. Of course, the drawback would be that more symbols need to be used to transmit the same amount of information, so the rate decreases.

We now make the above considerations more rigorous. We start by modelling the effect of the noise as inducing a conditional distribution on the received symbols given the transmitted symbols, and define the probability of error as the maximum over $i \in \{1, \ldots, M\}$ of the probabilities, that given index i was sent, the decoding function fails to identify it. We then define an *achievable rate* as follows.

Definition 5.1.2 *A rate $R > 0$ is achievable if for all $\epsilon > 0$ and for m large enough, there exists an encoding function $X^m : \{1, 2, \ldots, M\} \longrightarrow S^m$ and a decoding function $g : S^m \longrightarrow \{1, 2, \ldots, M\}$ over blocks of m symbols, whose rate is R and whose probability of error is less than ϵ.*

A key point of the above definition is that to achieve a certain rate, the probability of error is made arbitrarily small by encoding over larger block lengths. In other words, the random effect of the noise becomes negligible by taking larger and larger codewords.

We now ask what rates are achievable on a given channel. Notice that at this point is not even clear whether non-zero rates can be achieved at all. It may very well be the case that as $\epsilon \to 0$, the amount of redundancy we need to add to the codeword to combat the noise drives the rate to zero. Indeed, in the early days of communication theory it was believed that the only way to decrease the probability of error was to proportionally reduce the rate. A striking result of Shannon (1948) showed the belief to be incorrect: by accurately choosing encoding and decoding functions, one can communicate at a strictly positive rate, and at the same time with as small probability of error as desired. In other words, as $m \to \infty$, it is possible to transfer an amount of information $\log M$ of order at least m, and at the same time ensure that the probability of error is below ϵ. Shannon also showed that there is a highest achievable *critical rate*, called the *capacity* of the channel, for which this can be done. If one attempts to communicate at rates above the channel capacity, then it is impossible to do so with asymptotically vanishing error probability. Next, we formally define Shannon's capacity and explicitly determine it for a specific (and practical) channel model.

Definition 5.1.3 *The capacity of the channel is the supremum of the achievable rates, over all possible codeword distributions subject to a mean square constraint.*

It should be emphasised that the actual value of the capacity depends on both the noise model, and on the value of the codeword mean square constraint β given by (5.5).

5.1.2 Additive Gaussian channel

We consider a channel where both transmitted and received symbols take continuous values in $\mathbb{R}$. As far as the noise is concerned, in real systems this is due to a variety of causes. The cumulative effect can often be modelled as a Gaussian random variable that is added independently to each transmitted symbol, so that for a transmitted random codeword $X^m = (X_1, \ldots, X_m) \in \mathbb{R}^m$, the corresponding received codeword $Y^n = (Y_1, \ldots, Y_m) \in \mathbb{R}^m$ is obtained by

$$Y_i = X_i + Z_i \quad i = 1, 2, \ldots, m, \tag{5.6}$$

where the Z_i are i.i.d. Gaussian, mean zero, variance σ^2, random variables. A schematic representation of this Gaussian channel is given in Figure 5.2.

We have the following theorem due to Shannon (1948).

Theorem 5.1.4 *The capacity of the discrete time additive Gaussian noise channel $Y_i = X_i + Z_i$, subject to mean square constraint β and noise variance σ^2, is given by*

$$C = \frac{1}{2} \log \left(1 + \frac{\beta}{\sigma^2} \right) \text{ bits per symbol.} \tag{5.7}$$

Notice that in the capacity expression (5.7), the input constraint appears in terms of β, and the noise constraint appears in terms of σ^2. Not surprisingly, the capacity is larger

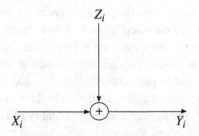

Fig. 5.2 The additive Gaussian channel.

as we relax the input constraint, by allowing larger values of β, or as we relax the noise constraint, by considering smaller values of σ^2.

A complete proof of Theorem 5.1.4 can be found in any information theory textbook and we shall not give it here. However, perhaps a little reflection can help to understand how the result arises.

Proof sketch of Theorem 5.1.4 The main idea is that the noise places a resolution limit on the possible codewords that can be distinguished at the receiver. Let us look at the constraints on the transmitted symbols, on the noise, and on the received symbols. By (5.5) we have that the typical range for the symbol values is $2\sqrt{\beta}$. Similarly, the typical range where we can observe most values of the noise is 2σ. Now, by looking at a codeword composed of m symbols, we have that

$$\sum_{i=1}^{m} E(Y_i^2) = \sum_{i=1}^{m} E((X_i + Z_i)^2)$$

$$= \sum_{i=1}^{m} E(X_i^2) + E(Z_i^2)$$

$$\leq m(\beta + \sigma^2). \tag{5.8}$$

It follows that the typical received codeword lies within a sphere of radius $\sqrt{m(\beta + \sigma^2)}$ in the space defined by the codeword symbols. In other words, the uncertainty due to the noise can be seen as a sphere of radius $\sqrt{m\sigma^2}$ placed around the transmitted codeword that lies within the sphere of radius $\sqrt{m\beta}$. Now, in order to recover different transmitted codewords without error, we want to space them sufficiently far apart so that their noise perturbed versions are still distinguishable. It turns out that the maximum number of codewords that can be reliably distinguished at the receiver is given by the maximum number of disjoint noise spheres of radius $\sqrt{m\sigma^2}$ that can be packed inside the received codeword sphere. This roughly corresponds to the ratio between the volume of the received codeword sphere and the volume of the noise sphere, see Figure 5.3, and is given by

$$\frac{\left(\sqrt{m(\beta + \sigma^2)}\right)^m}{(\sqrt{m}\sigma)^m} = \left(1 + \frac{\beta}{\sigma^2}\right)^{\frac{m}{2}}. \tag{5.9}$$

Clearly, this also corresponds to the number of resolvable elements of information that a codeword of length m can carry. The maximum number of resolvable bits per symbol is then obtained by taking the logarithm and dividing by m, yielding

$$\frac{1}{m} \log \left(1 + \frac{\beta}{\sigma^2}\right)^{\frac{m}{2}} = \frac{1}{2} \log \left(1 + \frac{\beta}{\sigma^2}\right), \tag{5.10}$$

which coincides with (5.7).

A more involved ergodic-theoretic argument also shows that the rate in (5.10) is achievable and this concludes the proof. The argument is based on a random code construction which roughly goes as follows. First, we randomly generate $M = 2^{mR}$ codewords of length m which are known to both transmitter and receiver that we call the codebook.

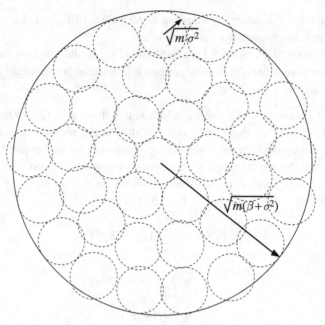

Fig. 5.3   The number of noise spheres that can be packed inside the received codeword sphere yields the maximum number of codewords that can be reliably distinguished at the receiver.

Then, one codeword is selected as the desired message, uniformly at random among the ones in the codebook, and it is sent over the channel. A noise corrupted version of this codeword is received. The receiver then selects the codeword of the codebook that is 'jointly typical' with the received one, if a unique jointly typical codeword exists, otherwise an error is declared. This roughly means that the receiver selects the codeword that is expected probabilistically to occur, given the received codeword and the statistics of the source and the channel. An ergodic-theoretic argument (the so-called asymptotic equipartition property) ensures that as m grows, the probability of not finding a jointly typical codeword, or of finding an incorrect one, tends (on average over all codebooks) to zero and concludes the proof sketch.   □

   It is interesting to note that in the above proof sketch, we never explicitly construct codes. Shannon's argument is based on the so-called probabilistic method, proving the existence of a code by showing that if one picks a random code, the error probability tends on average to zero, which means that there must exist a capacity achieving code. The engineering problem of designing such codes has been one of the major issues in information theory for many years. Clearly, simple repetition schemes such as 'send the message three times and use a two-out-of-three voting scheme if the received messages differ' are far from Shannon's limit. However, advanced techniques come much closer to reaching the theoretical limit, and today codes of rate essentially equal to Shannon's capacity, that are computationally practical and amenable to implementation have been invented, so we can safely assume in our network models that the actual rate of communication is given by (5.7).

5.1.3 Communication with continuous time signals

In real communication systems transmitted symbols are represented by continuous functions of time. For example, in communication over electrical lines transmitted symbols can be associated with variations of a voltage signal over time; in wireless communication with variations of the characteristics of the radiated electromagnetic field; in some biological systems with variations of the concentration of a certain chemical that is spread by a cell. Accordingly, to better model real communication systems, we now introduce a continuous version of the additive Gaussian noise channel.

Continuous physical signals are subject to two natural limitations: (i) their variation is typically bounded by power constraints, and (ii) their rate of variation is also limited by the medium where the signals propagate. These limitations, together with the noise process, place a fundamental limit on the amount of information that can be communicated between transmitter and receiver.

We start by letting a signal $x(t)$ be a continuous function of time, we define the instantaneous signal power at time t as $x^2(t)$, and the energy of the signal in the interval $[0, T]$ as $\int_0^T x^2(t)dt$. Let us now assume that we wish to transmit a specific codeword $X^m = (x_1, x_2, \ldots, x_m)$ over a perfect channel without noise. To do this, we need to convert the codeword into a continuous function of time $x(t)$ and transmit it over a suitable transmission interval $[0, T]$. One way to do this is to find a set of orthonormal functions $\phi_i(t)$, $i = 1, 2, \ldots, m$ over $[0, T]$, that is, functions satisfying

$$\int_0^T \phi_i(t)\phi_j(t)dt = \begin{cases} 1 & \text{if } i = j, \\ 0 & \text{if } i \neq j, \end{cases} \tag{5.11}$$

and transmit the signal

$$x(t) = \sum_{i=1}^m x_i\phi_i(t) \tag{5.12}$$

over that interval. The x_i can be recovered by integration:

$$x_i = \int_0^T x(t)\phi_i(t)dt. \tag{5.13}$$

According to the above strategy, a codeword X^m can be transmitted as a continuous signal of time $x(t)$ over an interval of length T. There is a limitation, however, on the possible signals $x(t)$ that can be transmitted. The restriction is that the transmitter cannot generate a signal of arbitrarily large power, meaning that $x^2(t)$ cannot exceed a maximum value P. This physical constraint is translated into an equivalent constraint on the codeword symbols x_i. First, notice that the total energy spent for transmission is limited by

$$\int_0^T x^2(t)dt \leq PT. \tag{5.14}$$

This leads to the following constraint on the sum of the codeword symbols:

$$\sum_{i=1}^{m} x_i^2 = \sum_{i=1}^{m} x_i \sum_{j=1}^{m} x_j \int_0^T \phi_i(t)\phi_j(t)\,dt$$

$$= \int_0^T \sum_{i=1}^{m} x_i\phi_i(t) \sum_{j=1}^{m} x_j\phi_j(t)\,dt$$

$$= \int_0^T x^2(t)\,dt$$

$$\leq PT, \tag{5.15}$$

where the equalities follow from the signal representation (5.12), the orthonormality condition (5.11), and the last inequality from (5.14).

We have found that the physical power constraint imposes a constraint on the codeword symbols which, letting $PT/m = \beta$, we can rewrite in terms of the mean square constraint,

$$\frac{1}{m}\sum_{i=1}^{m} x_i^2 \leq \beta. \tag{5.16}$$

Notice that above constraint is very similar to (5.5). For example, assuming the input symbols to be ergodic, we have that

$$\lim_{m\to\infty} \frac{1}{m}\sum_{i=1}^{m} x_i^2 = E(X_1^2). \tag{5.17}$$

However, we also notice that while in (5.5) β is a given constant, in (5.16) β depends on m, T, and on the constant P. We expect that in practice one cannot transmit an arbitrary number m of symbols in a finite time interval T, and that for physical reasons T must be proportional to m, so that β stays constant also in this case.

It turns out that every physical channel has exactly this limitation. Every channel is characterised by a constant *bandwidth* $W = m/2T$ that limits the amount of variation over time of any signal that is transmitted through it. This means that there is a limit on the number of orthonormal basis functions that can represent a signal $x(t)$, when such a signal is sent over the channel. This limits the amount of diversity, or degrees of freedom of the signal, in the sense that if $x(t)$ is sent over time T, it can be used to distinguish at most among $m = 2WT$ different symbols. This is known as the *Nyquist number*, after Nyquist (1924). If one tries to transmit a number of symbols above the Nyquist number, the corresponding signal will appear distorted at the receiver and not all the symbols can be recovered. In other words, one can think of the channel as a filter limiting the amount of diversity of the signal that is transmitted through it.

We can now put things together and consider the act of communicating a random signal $X(t)$ in the interval $[0, T]$ and in the presence of noise. Accordingly, we consider a continuous random noise process $Z(t)$ added to the signal $X(t)$, so that

$$Y(t) = X(t) + Z(t). \tag{5.18}$$

We model the noise process as *white Gaussian*, which for us simply means that $Z_i = \int_0^T Z(t)\phi_i(t)dt$ are i.i.d. Gaussian, mean zero, variance σ^2, random variables. Thus, when the receiver attempts to recover the value of X_i by integration according to (5.13) we have,

$$Y_i = \int_0^T (X(t) + Z(t))\phi_i(t)dt = X_i + Z_i, \qquad (5.19)$$

and we are back to the discrete-time channel representation depicted in Figure 5.2, subject to the mean square constraint, for which Shannon's Theorem 5.1.4 applies. Theorem 5.1.4 gives an expression of the capacity in bits per symbol. We have seen that m symbols are transmitted in a period of T seconds using a continuous function of time, and that $\beta = PT/m = P/2W$. Letting $\sigma^2 = N/2$, and expressing the capacity in bits per second rather than in bits per symbol, we immediately have the following theorem that is known as the Shannon–Hartley theorem.

Theorem 5.1.5 *The capacity of the continuous time, additive white Gaussian noise channel $Y(t) = X(t) + Z(t)$, subject to the power constraint $x^2(t) \leq P$, bandwidth $W = m/2T$, and noise spectral density $N = 2\sigma^2$, is given by*

$$C = W \log\left(1 + \frac{P}{NW}\right) \ bits \ per \ second. \qquad (5.20)$$

5.1.4 Information-theoretic random networks

We now turn our attention to random networks. Let us consider points of a planar Poisson point process X of density one inside a box B_n of size $\sqrt{n} \times \sqrt{n}$. We start by describing the act of communication between a single pair of points, which is governed by the the the Shannon–Hartley formula (5.20).

Let us denote by $x(t)$ the signal transmitted by node $x \in X$ over a time interval T, and by $y(t)$ the corresponding signal received by node $y \in X$. Furthermore, let us assume that the transmitted signal is subject to a loss factor $\ell(x, y)$, which is a function from $\mathbb{R}^2 \to \mathbb{R}^+$. Node y then receives the signal

$$y(t) = x(t)\ell(x, y) + z(t), \qquad (5.21)$$

where $z(t)$ is a realisation of the white Gaussian noise process $Z(t)$. Given, as before, an instantaneous power constraint on the transmitting node $x^2(t) < P$ for all t, we immediately have the constraint $x^2(t)\ell^2(x, y) < P\ell^2(x, y)$. From Theorem 5.1.5 it then follows, assuming unit bandwidth, that when x attempts to communicate with y the capacity is given by

$$C(x, y) = \log\left(1 + \frac{P\ell^2(x, y)}{N}\right). \qquad (5.22)$$

Next, we assume that the loss $\ell(x, y)$ only depends on the Euclidean norm $|x - y|$ and is decreasing in the norm. Hence, we have $\ell(x, y) = l(|x - y|)$ for some decreasing function $l : \mathbb{R}^+ \to \mathbb{R}^+$, such that it satisfies the integrability condition $\int_{\mathbb{R}^+} xl^2(x)dx < \infty$. Notice now that if we select uniformly at random two points $x, y \in X$ inside B_n, their average distance is of the order $\sqrt{n}$. Since $\ell(\cdot)$ is a decreasing function of the distance between transmitter and receiver, it is a simple exercise to show that for all $\epsilon > 0$, when x attempts to communicate with y,

$$\lim_{n \to \infty} P(C(x, y) > \epsilon) = 0. \tag{5.23}$$

It is natural to ask whether, by also allowing the other nodes in B_n to transmit, it is possible to design a collective strategy that allows us to have a non-zero rate between x and y. We notice that if we can find a chain of nodes between x and y, such that the distance between any two consecutive nodes along the chain is bounded above by a constant, then by (5.22) the capacity between each pair of nodes along the chain is non-zero, and these nodes can be used as successive relays for communication between x and y. It follows that x and y can in principle achieve a strictly positive rate. In this case however, we have to account for an additional term when we model the communication channel, that is the interference due to simultaneous transmissions along the chain. Let x_i and x_j be two successive nodes along the chain that act as relays for the communication between x and y, and let $\mathcal{C}$ be the set of nodes in the chain that are simultaneously transmitting when x_i transmits to x_j. The signal received by x_j is then given by

$$x_j(t) = x_i(t)\ell(x_i, x_j) + z(t) + \sum_{x_k \in \mathcal{C}: k \neq i} x_k(t)\ell(x_k, x_j), \tag{5.24}$$

where the sum accounts for all interfering nodes along the chain. Clearly, x_j is only interested in decoding the signal from x_i, and if we treat all the interference as Gaussian noise, we have that an achievable rate between x_i and x_j is given by

$$R(x_i, x_j) = \log\left(1 + \frac{P\ell^2(x_i, x_j)}{N + P\sum_{x_k \in \mathcal{C}: k \neq i} \ell^2(x_k, x_j)}\right). \tag{5.25}$$

Now, if in order to reduce the interference we let only one out of every k nodes along the chain transmit simultaneously, and we show that in this case $R(x_i, x_j) \geq R > 0$ for all of them, it then follows that an achievable rate between the source x and the final destination y is given by the ratio of R/k. It should be emphasised that this rate would only be a *lower bound* on the capacity between x and y, obtained by performing a specific relay scheme that treats interference on the same footing as random noise, and performs pairwise point-to-point coding and decoding along a chain of nodes. In principle, the capacity can be higher if one constructs more complex communication schemes. For example, to achieve a higher rate, if there are some dominant terms in the interference sum in (5.25), rather than treat them as noise, x_j could try first to decode them, then subtract them from the received signal, and finally decode the desired signal. Instead, an *upper bound* on the capacity should be independent of any communication scheme.

In the next section, we start by considering in more detail the case outlined above when only two nodes in the network wish to communicate. We show that there exists a set S_n containing an arbitrarily large fraction α of the nodes, such that we can choose any pair of nodes inside S_n and have w.h.p. a positive rate $R(\alpha)$ between them. To this end, we construct a relay scheme of communication along a chain of nodes. For this scheme the rate $R(\alpha)$ tends to zero as $\alpha \to 1$. One might then think that this is a limitation of the specific scheme used. However, it turns out that *regardless of the scheme used for communication*, there always exists a set $\overline{S}_n$ containing at least a positive fraction $(1 - \alpha)$ of the nodes, such that the largest achievable rate among any two nodes inside $\overline{S}_n$ is zero w.h.p. When viewing these results together, we conclude that it is not possible to have *full-information connectivity* inside B_n, but it is possible to have *almost-full information connectivity*.

We then move to the case when many pairs of nodes wish to communicate between each other simultaneously, and compute upper and lower bounds on the achievable per-node rate. We show an inverse "square-root law" on the achievable per-node rate, as the number of nodes that are required to communicate increases.

5.2 Scaling limits; single source–destination pair

We now formalise the discussion for the single source–destination pair outlined in the previous section. We assume that only two nodes in the network wish to communicate. We first show that almost all pairs of nodes in the network can communicate at a constant rate. This result follows from the almost connectivity property of the boolean model, see Theorem 3.3.3, and is based on the simple construction described before, where pairs of nodes communicate using a chain of successive transmissions, in a multi-hop fashion. Each transmission along the chain is performed at rate given by (5.25), and we design an appropriate time schedule for transmissions along the path. The second result we show is slightly more involved and it requires some additional information-theoretic tools for its proof. It rules out the possibility that all pairs of nodes in the network can communicate at a constant rate, *irrespective* of their cooperation strategies. In this case, nodes are not restricted to pairwise coding and decoding, but are allowed to jointly cooperate in arbitrarily complex ways. Fortunately, as we shall see, information theory provides a basic cut-set argument that allows us to bound the total information flow between any two sides of the network, and this will be the key to show that the rate must go to zero for at least a small fraction of the nodes.

Theorem 5.2.1 *For any $R > 0$ and $0 < \alpha < 1$, let $A_n(R, \alpha)$ be the event that there exists a set S_n of at least αn nodes, such that for all $s, d \in S_n$, s can communicate with d at rate R. Then for all $0 < \alpha < 1$ there exists $R = R(\alpha) > 0$, independent of n, such that*

$$\lim_{n \to \infty} P(A_n(R(\alpha), \alpha)) = 1. \tag{5.26}$$

Proof Consider the boolean model network of density one and radius r inside B_n. By Theorem 3.3.3 it follows that for any $0 < \alpha < 1$ it is possible to choose r large enough so that the network is α-almost connected w.h.p. Let S_n be the largest connected component of the boolean model inside B_n, and let us pick any two nodes $s, d \in S_n$. Consider the shortest path connecting s to d. We shall compute an achievable rate along this path. Let us first focus on a pair (x_i, x_j) of neighbouring nodes along the path, where x_i is the transmitter and x_j is the receiver. Clearly, these nodes are at a distance of at most $2r$ from each other. By (5.25) and since $l(\cdot)$ is a decreasing function, we have

$$R(x_i, x_j) = \log\left(1 + \frac{P\ell^2(x_i, x_j)}{N + P\sum_{x_k \in \mathcal{C}: k \neq i} \ell^2(x_k, x_j)}\right)$$

$$\geq \log\left(1 + \frac{Pl^2(2r)}{N + P\sum_{x_k \in \mathcal{C}: k \neq i} l^2(x_k - x_j)}\right). \tag{5.27}$$

We now make a geometric observation to bound the interference term in the denominator of (5.27). With reference to Figure 5.4, we observe that any ball along the shortest path can only overlap with its predecessor and its successor. Otherwise, if it had overlapped with any other ball, it would have created a shortcut, which would have avoided at least one other ball and thus give a shorter path, which is impossible.

With this observation in mind, we divide time into slots, and in each time slot we allow only one out of three nodes along the path to transmit a signal. In this way, we guarantee that during a time slot, each receiver has its predecessor transmitting, but not its successor, and all nodes transmitting in a given time slot are at distance more than $2r$ from all receiving nodes (except their intended relay), and from each other.

The interference term is then bounded by packing the transmitting balls in a 'honeycomb' lattice arrangement; see Figure 5.5. Note that we can partition this lattice into groups of $6k$ nodes, $k = 1, 2, \ldots$, lying on concentric hexagons, and whose distances from y are larger than kr. Hence, we have

$$\sum_{x_k \in \mathcal{C}: k \neq i} l^2(x_k - x_j) \leq \sum_{k=1}^{\infty} 6k l^2(kr)$$

$$= K(r), \tag{5.28}$$

where $K(r) < \infty$ since $\int_{\mathbb{R}^+} x l^2(x) dx < \infty$.

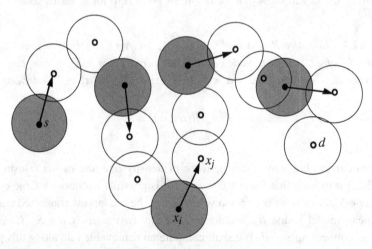

Fig. 5.4 Nodes on the shortest path from x to y that simultaneously transmit are further than $2r$ from any receiving node, except their intended relay, and from each other.

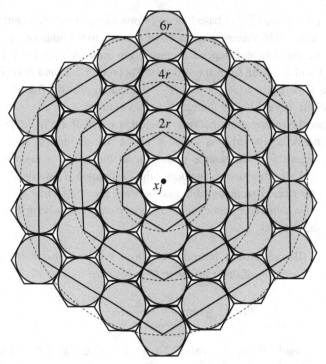

Fig. 5.5 Since balls interfering with *y* cannot overlap, the interference is bounded by the configuration where nodes are placed on a honey-comb lattice.

Substituting (5.28) into (5.27) we have that the following rate is achievable for each transmitter–receiver pair x_i, x_j along the path in every time slot,

$$R(x_i, x_j) \geq \log\left(1 + \frac{Pl^2(2r)}{N + PK(r)}\right). \tag{5.29}$$

Notice that $R(x_i, x_j)$ depends on α because this determines the value of r. As we used three time slots, an achievable rate for all $s, d \in S_n$ is $R(\alpha) = R(x_i, y_j)/3$, which completes the proof. □

Notice that in the proof of Theorem 5.2.1, as $\alpha \to 1$, the rate in (5.29) tends to zero, since $r \to \infty$. Of course, this does not rule out that in principle a different strategy could achieve a constant rate w.h.p. when $\alpha = 1$. A converse result however, shows that this is impossible and a non-vanishing rate cannot be achieved by all the nodes.

Theorem 5.2.2 *For any $R > 0$ and $0 < \alpha < 1$, let $\tilde{A}_n(R, \alpha)$ be the event that there exists a set S_n of at least αn nodes, such that for any $x, y \in S_n$, x cannot communicate with y at rate R. Then for any $R > 0$, there exists $\alpha(R) > 0$, such that*

$$\lim_{n \to \infty} P(\tilde{A}_n(R, \alpha(R))) = 1. \tag{5.30}$$

The proof of Theorem 5.2.2 is based on an information cut-set argument that provides a bound on the achievable information rate from one side to the other of a communication network. This bound does not depend on any given strategy used for communication. Furthermore, it will be clear from the proof of Theorem 5.2.2 (and it is indeed a good exercise to check this), that $\alpha(R) \to 0$ as $R \to 0$, meaning that the fraction of nodes that cannot achieve rate R vanishes as the rate becomes smaller.

Next, we state the information-theoretic result that we shall use in the proof. Consider an arbitrary network composed of a set U of nodes. We divide this set into two parts, as depicted in Figure 5.6. The source node s is on the left of a manifold named the *broadcast cut*, and all other nodes, including the destination d, are on the right of this cut. We denote by $R(s, x)$ an achievable rate between the source and a node $x \in U$ that can be obtained by some communication strategy, possibly involving transmissions by other nodes in the network.

Theorem 5.2.3   (Broadcast-cut bound) *For any $s \in U$ we have that the sum of the achievable rates from s to all other nodes in U is bounded as*

$$\sum_{x \in U, x \neq s} R(s, x) \leq \log \left(1 + \frac{P \sum_{x \in U, x \neq s} \ell^2(s, x)}{N} \right). \tag{5.31}$$

We make some remarks about Theorem 5.2.3. First, notice that if the number of nodes in the network is $|U| = 2$, then the source can only communicate directly with a single destination and (5.31) reduces to (5.22), which is the capacity of a single point to point communication. When $|U| > 2$, because of the presence of additional nodes, we might reasonably expect that we can use these additional nodes in some clever way to obtain individual point to point rates $R(s, x)$ higher than what (5.22) predicts for the single pair scenario, and Theorem 5.2.3 provides a bound on the sum of these rates. A crude bound on the individual rates can then be obtained by using the whole sum to bound its individual components, leading to the following corollary.

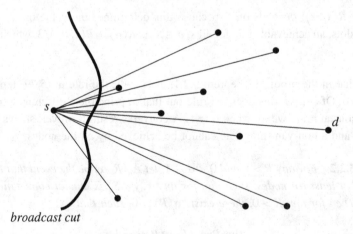

Fig. 5.6   The broadcast cut.

Corollary 5.2.4 *For all s, d ∈ U, the achievable rate between them is upper bounded as*

$$R(s, d) \leq \log\left(1 + \frac{P\sum_{x \in U, x \neq s} \ell^2(s, x)}{N}\right). \tag{5.32}$$

With this corollary in mind, we are now ready to proceed with the proof of Theorem 5.2.2.

Proof of Theorem 5.2.2 Consider two Poisson points s, d inside B_n. By Corollary 5.2.4 applied to the set $X \cap B_n$ of Poisson points falling inside B_n, we have that the sum of losses from s to all other Poisson points x inside the box is lower bounded as follows:

$$I(x) \equiv \sum_{x \in X \cap B_n, x \neq s} \ell^2(s, x) \geq \frac{N}{P}(2^{R(s,d)} - 1), \tag{5.33}$$

where (5.33) has been obtained by inversion of (5.32). Notice that (5.33) gives a necessary condition to achieve a constant rate $R(s, d)$ between any two nodes s and d. We now claim that w.h.p. this necessary condition does not hold for a positive fraction of the nodes in B_n. To see why this is, consider for any constant K the event $I(x) < K$. By integrability of the function $\ell^2(\cdot)$, we a.s. have

$$\sum_{i:|x_i - x| > L} \ell^2(x, x_i) \to 0, \tag{5.34}$$

as $L \to \infty$. Since almost sure convergence implies convergence in probability, it follows that we can choose $L(K)$ large enough such that

$$P\left(\sum_{i:|x_i - x| > L} \ell^2(x, x_i) < K\right) > \epsilon. \tag{5.35}$$

Now, notice that for any fixed L, there is also a positive probability such that no Poisson point lies inside the disc of radius L centred at x. This latter event is clearly independent of the one in (5.35), and considering them jointly we have

$$P(I(x) < K) > 0. \tag{5.36}$$

Finally, choose $K = N(2^R - 1)/P$ and let Y_n be the number of Poisson points inside the box B_n for which $I(x) < K$. By the ergodic theorem we have a.s.

$$\lim_{n \to \infty} \frac{Y_n}{n} = P(I(x) < K) > 0, \tag{5.37}$$

and the proof is complete. □

The proof of Theorem 5.2.2 was based on a necessary condition of having constant rate R based on Corollary 5.2.4: the sum of the losses from point x to all other points of the Poisson process must be larger than a certain constant. We now want to give a geometric interpretation of this result. Assuming the attenuation function to be symmetric, the sum of the losses from x can also be seen as the amount of interference that all Poisson points generate at x. The necessary condition can then be seen as having high enough interference $I(x)$ at any point inside B_n. Interference, i.e., the amount of signal received, becomes an essential ingredient

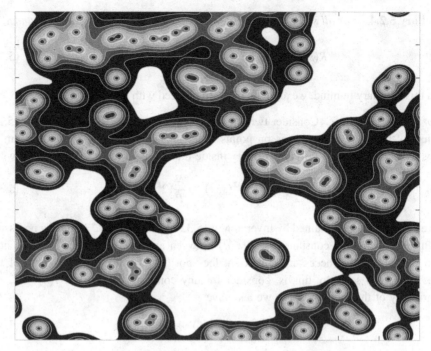

Fig. 5.7   Contour plot of the shot-noise process $I(x)$.

for communication among the nodes. Figure 5.7 illustrates this concept by showing the contour plot of the shot-noise process $I(x)$ inside the box B_n. The white colour represents the region where the necessary condition is not satisfied, i.e. any point placed inside the white region is isolated, in the sense that it cannot transmit at constant rate, because the value of the shot-noise is too low.

5.3 Multiple source–destination pairs; lower bound

So far we have considered a random network in which only two nodes wish to exchange information. In this section we consider a different scenario in which many pairs of nodes wish to communicate between each other simultaneously. More precisely, we pick uniformly at random a matching of source–destination pairs, so that each node is the destination of exactly one source. We shall determine an 'inverse square-root law' on the achievable per-node rate, as the number of nodes that are required to communicate increases.

We start by showing a lower bound on the achievable rate by explicitly describing a communication strategy that achieves the desired bound. Then, in the next section we derive an upper bound that is independent of any communication strategy. In the following, we assume a specific form of the attenuation function. Denoting the Euclidean distance between two nodes x_i and x_j by $|x_i - x_j|$, we assume attenuation of the power to be of the type $\ell^2(x_i, x_j) = e^{-\gamma|x_i - x_j|}$, with $\gamma > 0$, that is, we assume exponential decay of the power. It should be clear from the proof, and it is left as an exercise to check this, that by similar computations the same lower bound holds assuming a power attenuation function of the type $\ell^2(x_i, x_j) = \min\{1, |x_i - x_j|^\alpha\}$, with $\alpha > 2$. Attenuation functions such as the ones above are often used to model communications occurring in media with absorption, where typically exponential attenuation prevails.

In this section we use the following probabilistic version of the order notation as described in Appendix A.1. For positive random variables X_n and Y_n, we write $X_n = O(Y_n)$ w.h.p. if there exists a constant $K > 0$ such that

$$\lim_{n \to \infty} P(X_n \leq KY_n) = 1. \tag{5.38}$$

We also write $f(n) = \Omega(g(n))$, as $n \to \infty$, if $g(n) = O(f(n))$ in the sense indicated above. The main result of this section is the following.

Theorem 5.3.1 *W.h.p. all nodes in B_n can (simultaneously) transmit to their intended destinations at rate*

$$R(n) = \Omega(1/\sqrt{n}). \tag{5.39}$$

We now give an overview of the strategy used to achieve the above bound. As in the single source–destination pair scenario described in the previous section, we rely on multi-hop routing across percolation paths. This means that we divide nodes into sets that cross the network area. These sets form a 'highway system' of nodes that can carry information across the network at constant rate, using short hops. The rest of the nodes access the highway using single hops of longer length. The communication strategy is then divided into four consecutive phases. In a first phase, nodes drain their information to the highway, in a second phase information is carried horizontally across the network through the highway, in a third phase it is carried vertically, and in a last phase information is delivered from the highway to the destination nodes. Figure 5.8 shows a schematic representation of the first phase. In each phase we use point-to-point coding and decoding on each Gaussian channel between transmitters and receivers, and design an appropriate time schedule for transmission.

Given the construction outlined above, and letting all nodes transmit with the same power constraint P, we might expect the longer hops needed in the first and last phases of the strategy to have a lower bit-rate, due to higher power loss across longer distances. However, one needs to take into account other components that influence the bit-rate, namely, interference, and relay of information from other nodes. It turns out that when all these components are accounted for, the bottleneck is due to the information carried through the highway.

We now give a short sketch of the proof. First, we notice that the highway consists of paths of hops whose length is uniformly bounded above by some constant. Then, using a time division protocol similar to the one described in the proof of Theorem 5.2.1, we show that a constant transmission rate can be achieved along each path. However, we also need to account for the relay of information coming from all nodes that access a given path. The number of these nodes is at most proportional to $\sqrt{n}$. This is ensured

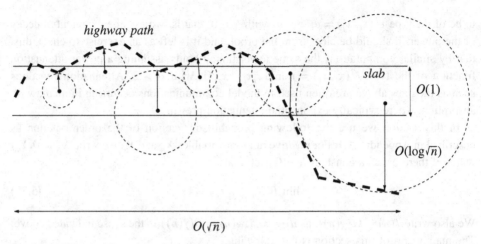

Fig. 5.8 Nodes inside a slab of constant width access a path of the highway in single hops of length at most proportional to $\log \sqrt{n}$. Multiple hops along the highway are of constant length.

by associating with each path only those nodes that are within a slab of constant width that crosses the network area; see Figure 5.8. It follows that the rate of communication of each node on the highway paths can be of order $1/\sqrt{n}$.

Now, let us consider the rate of the nodes that access the highway in single hops. The proof is completed by showing that these nodes, requiring single hops of length at most proportional to $\log \sqrt{n}$, and not having any relay burden, can sustain a rate higher than $1/\sqrt{n}$.

Notice that there are three key points in our reasonings: (i) there exist paths of constant hop length that cross the entire network forming the highway system, (ii) these paths can be put into a one-to-one correspondence with $\sqrt{n}$ slabs of constant width, each containing at most a constant times $\sqrt{n}$ number of nodes, and (iii) these paths are somehow regularly spaced so that there is always one within a $\log \sqrt{n}$ distance factor from any node in the network.

In the following, a mapping to a discrete percolation model allows application of Theorem 4.3.9 to ensure the existence of many crossing paths. A time division strategy, in conjunction with a counting argument, shows that each highway path can have a constant rate, and that nodes can access the highway at a rate at least proportional to $1/\sqrt{n}$. Finally, some simple concentration bounds show that the number of nodes that access any given path is at most a constant times $\sqrt{n}$.

5.3.1 The highway

To begin our construction, we partition the box B_n into subsquares s_i of constant side length c, as depicted in the left-hand of Figure 5.9. Let $X(s_i)$ be the number of Poisson points inside s_i. By choosing c, appropriately we can arrange that the probability that a square contains at least a Poisson point is as high as we want. Indeed, for all i, we have

$$p \equiv P(X(s_i) \geq 1) = 1 - e^{-c^2}. \tag{5.40}$$

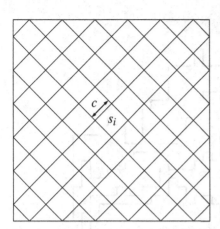 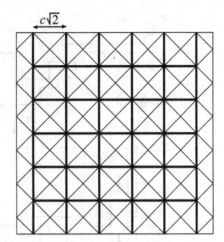

Fig. 5.9 Construction of the bond percolation model. We declare each square on the left-hand side of the picture open if there is at least a Poisson point inside it, and closed otherwise. This corresponds to associating an edge with each square, traversing it diagonally, as depicted on the right-hand side of the figure, and declaring the edge either open or closed according to the state of the corresponding square.

We say that a square is *open* if it contains at least one point, and *closed* otherwise. Notice that squares are open with probability p, independently of each other.

We now map this model into a discrete bond percolation model on the square grid. We draw a horizontal edge across half of the squares, and a vertical edge across the others, as shown on the right-hand side of Figure 5.9. In this way we obtain a grid G_n of horizontal and vertical edges, each edge being open, independently of all other edges, with probability p. We call a path of G_n *open* if it contains only open edges. Note that, for c large enough, by Theorem 4.3.8, our construction produces open paths that cross the network area w.h.p.; see Figure 5.10 for a simulation of this. It is convenient at this point to denote the number of edges composing the side length of the box by $m = \sqrt{n}/(c\sqrt{2})$, where c is rounded up such that m is an integer. Recall from Theorems 4.3.8 and 4.3.9 that, as shown in Figure 4.8, there are w.h.p. $\Omega(m)$ paths in the whole network, and that these can be grouped into disjoint sets of $\lceil \delta \log m \rceil$ paths, each group crossing a rectangle of size $m \times (\kappa \log m - \epsilon_m)$, by appropriately choosing κ and δ, and a vanishingly small ϵ_m so that the side length of each rectangle is an integer. The same is true of course if we divide the area into vertical rectangles and look for paths crossing the area from bottom to top. Using the union bound, we conclude that there exist both horizontal and vertical disjoint paths w.h.p. These paths form a backbone, that we call the *highway system*.

5.3.2 Capacity of the highway

Along the paths of the highway system, we choose one Poisson point per edge, that acts as a relay. This is possible as the paths are formed by open edges, which are associated with non-empty squares. The paths are thus made of a chain of nodes such that the distance between any two consecutive nodes is at most $2\sqrt{2}c$.

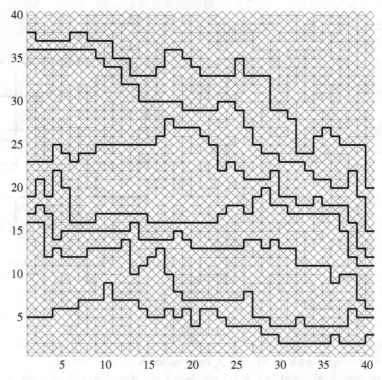

Fig. 5.10   Horizontal paths in a 40×40 bond percolation model obtained by computer simulation. Each square is traversed by an open edge with probability p ($p = 0.7$ here). Closed edges are not depicted.

To achieve a constant rate along a path, we now divide time into slots. The idea is that when a node along a path transmits, other nodes that are sufficiently far away can simultaneously transmit, without causing excessive interference. The following theorem makes this precise, ensuring that a constant rate R, independent of n, can be achieved w.h.p. on all the paths simultaneously. The theorem is stated in slightly more general terms considering nodes at L_1 distance d in the edge percolation grid G_n, rather than simply neighbours, as this will be useful again later. Notice that the rate along a crossing path can be obtained immediately by letting $d = 1$.

Theorem 5.3.2   *For any integer $d > 0$, there exists an $R(d) > 0$, such that in each square s_i there is a node that can transmit w.h.p. at rate $R(d)$ to any destination located within distance d. Furthermore, as d tends to infinity, we have*

$$R(d) = \Omega \left(d^{-2} e^{-\gamma \sqrt{2} cd} \right). \tag{5.41}$$

Proof   We divide time into a sequence of k^2 successive slots, with $k = 2(d + 1)$. Then, we consider disjoint sets of subsquares s_i that are allowed to transmit simultaneously, as depicted in Figure 5.11.

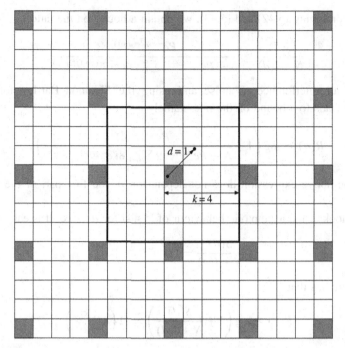

Fig. 5.11 The situation depicted represents the case $d = 1$. Grey squares can transmit simultaneously. Notice that around each grey square there is a 'silence' region of squares that are not allowed to transmit in the given time slot.

Let us focus on one given subsquare s_i. The transmitter in s_i transmits towards a destination located in a square at distance at most d (diagonal) subsquares away. First, we find an upper bound for the interference at the receiver. We notice that the transmitters in the 8 closest subsquares are located at Euclidean distance at least $c(d+1)$ from the receiver; see Figure 5.11. The 16 next closest subsquares are at Euclidean distance at least $c(3d+3)$, and so on. By extending the sum of the interferences to the whole plane, this can then be bounded as

$$I(d) \leq \sum_{i=1}^{\infty} 8i \, Pl(c(2i-1)(d+1))$$

$$= Pe^{-\gamma c(d+1)} \sum_{i=1}^{\infty} 8i \, e^{-\gamma c(d+1)(2i-2)}; \qquad (5.42)$$

notice that this sum converges if $\gamma > 0$.

Next, we want to bound from below the signal received from the transmitter. We observe first that the distance between the transmitter and the receiver is at most $\sqrt{2}c(d+1)$. Hence, the signal $S(d)$ at the receiver can be bounded by

$$S(d) \geq Pl(\sqrt{2}c(d+1))$$

$$= Pe^{-\gamma\sqrt{2}c(d+1)}. \qquad (5.43)$$

Finally, by combining (5.42) and (5.43), we obtain a bound on the ratio,

$$\frac{S(d)}{N+I(d)} \geq \frac{Pe^{-\gamma\sqrt{2}c(d+1)}}{N+Pe^{-\gamma c(d+1)}\sum_{i=1}^{\infty} 8i\, e^{-\gamma c(d+1)(2i-2)}}. \qquad (5.44)$$

By treating all interference as noise and using the Shannon–Hartley formula, this immediately leads to a bound on the rate, namely

$$R(d) \geq \log\left(1 + \frac{Pe^{-\gamma\sqrt{2}c(d+1)}}{N+Pe^{-\gamma c(d+1)}\sum_{i=1}^{\infty} 8i\, e^{-\gamma c(d+1)(2i-2)}}\right), \qquad (5.45)$$

and since the above expression does not depend on n, the first part of the theorem is proven.

We now look at the asymptotic behaviour of (5.44) for $d \to \infty$. It is easy to see that

$$\frac{S(d)}{N+I(d)} = \Omega\left(e^{-\gamma\sqrt{2}cd}\right), \qquad (5.46)$$

which also implies that

$$R(d) \geq \log\left(1 + \frac{S(d)}{N+I(d)}\right) = \Omega\left(e^{-\gamma\sqrt{2}cd}\right). \qquad (5.47)$$

Finally, accounting for the time division into $k^2 = 4(d+1)^2$ time slots, the actual rate available in each square is $\Omega(d^{-2}e^{-\gamma\sqrt{2}cd})$. $\qquad\square$

The proof of the following corollary is immediate by switching the role of transmitters and receivers in the above proof. Distances remain the same, and all equations still hold.

Corollary 5.3.3 *For any integer $d > 0$, there exists an $R(d) > 0$, such that in each square s_i there is a node that can can receive w.h.p. at rate $R(d)$ from any transmitter located within distance d. Furthermore, as d tends to infinity, we have w.h.p.*

$$R(d) = \Omega\left(d^{-2}e^{-\gamma\sqrt{2}cd}\right). \qquad (5.48)$$

5.3.3 Routing protocol

Given the results of the previous section, we can now describe a routing protocol that achieves $\Omega(1/\sqrt{n})$ per-node rate. The protocol uses four separate phases, and in each phase time is divided into slots. A first phase is used to drain information to the highway, a second one to transport information on the horizontal highways connecting the left and right edges of the domain, a third one to transport information on the vertical highways connecting the top and bottom edges of the domain, and a fourth one to deliver information to the destinations. The draining and delivery phases use direct transmission and multiple time slots, while the highway phases use both multiple hops and multiple time slots. We show that the communication bottleneck is in the highway phase which can achieve a per-node rate of $\Omega(1/\sqrt{n})$.

We start by proving two simple lemmas that will be useful in the computation of the rate.

Lemma 5.3.4 *If we partition B_n into an integer number $m^2 = n/c^2$ of subsquares s_i of constant side length c, then there are w.h.p. less than $\log m$ nodes in each subsquare.*

Proof The proof proceeds via Chernoff's bound in Appendix A.4.3. Let A_n be the event that there is at least one subsquare with more than $\log m$ nodes. Since the number of nodes $|s_i|$ in each subsquare of the partition is a Poisson random variable of parameter c^2, by the union and Chernoff bounds, we have

$$P(A_n) \leq m^2 P(|s_i| > \log m)$$

$$\leq m^2 e^{-c^2} \left(\frac{c^2 e}{\log m} \right)^{c^2 \log m}$$

$$= e^{-c^2} \left(\frac{c^2 e^{2/c^2 + 1}}{\log m} \right)^{c^2 \log m} \to 0, \qquad (5.49)$$

as m tends to infinity. $\qquad \square$

Lemma 5.3.5 *If we partition B_n into an integer number $\sqrt{n}/w$ of rectangles R_i of side lengths $\sqrt{n} \times w$, then there are w.h.p. less than $2w\sqrt{n}$ nodes in each rectangle.*

Proof Again, the proof proceeds via Chernoff's bound in Appendix A.4.3. Let A_n be the event that there is at least one rectangle with more than $2w\sqrt{n}$ nodes. Since the number of nodes $|R_i|$ in each rectangle is a Poisson random variable of parameter $w\sqrt{n}$, by the union and Chernoff bounds, we have

$$P(A_n) \leq \frac{\sqrt{n}}{w} P(|R_i| > 2w\sqrt{n})$$

$$\leq \frac{\sqrt{n}}{w} e^{-w\sqrt{n}} \left(\frac{ew\sqrt{n}}{2w\sqrt{n}} \right)^{2w\sqrt{n}}$$

$$= \frac{\sqrt{n}}{w} e^{-w\sqrt{n}} \left(\frac{e}{2} \right)^{2w\sqrt{n}} \to 0, \qquad (5.50)$$

as n tends to infinity. $\qquad \square$

The next lemma illustrates the achievable rate in the draining phase of the protocol, occurring in a single hop.

Lemma 5.3.6 *Every node inside B_n can w.h.p. achieve a rate to some node on the highway system of $\Omega \left((\log n)^{-3} n^{-\sqrt{2} c \kappa \gamma / 2} \right)$.*

Proof We want to compute an achievable rate from sources to the highways. Recall that $p = 1 - e^{-c^2}$. By Theorem 4.3.9 we can partition the square B_n into an integer number of rectangles of size $m \times (\kappa \log m - \epsilon_m)$ and choose κ and c such that there are at least

$\lceil \delta \log m \rceil$ crossing paths in each rectangle w.h.p. We then slice the network area into horizontal strips of constant width w, by choosing w appropriately such that there are at least as many paths as slices inside each rectangle of size $m \times (\kappa \log m - \epsilon_m)$. We can then impose that nodes from the ith slice communicate directly with the ith horizontal path. Note that each path may not be fully contained in its corresponding slice, but it may deviate from it. However, a path is never further than $\kappa \log m - \epsilon_m$ from its corresponding slice.

More precisely, to each source in the ith slab, we assign an *entry point* on the ith horizontal path. The entry point is defined as the node on the horizontal path closest to the vertical line drawn from the source point, see Figure 5.12. The source then transmits directly to the entry point. Theorem 4.3.9 and the triangle inequality ensure that the distance between sources and entry points is never larger than $\kappa \log m + \sqrt{2}c$. This is because each rectangle contains $\lceil \delta \log m \rceil$ paths, and therefore each source finds its highway within the same rectangle.

Hence, to compute the rate at which nodes can communicate to the entry points, we let $d = \kappa \log m + \sqrt{2}c$ and apply the second part of Theorem 5.3.2. We obtain that one node per square can communicate to its entry point at rate

$$R(\kappa \log m + \sqrt{2}c) = R\left(\kappa \log \frac{\sqrt{n}}{\sqrt{2}c} + \sqrt{2}c\right)$$

$$= \Omega\left(\frac{e^{-\gamma\sqrt{2}c\kappa \log \frac{\sqrt{n}}{\sqrt{2}c}}}{\left(\kappa \log \frac{\sqrt{n}}{\sqrt{2}c}\right)^2}\right)$$

$$= \Omega\left(\frac{n^{-\frac{\sqrt{2}}{2}c\kappa\gamma}}{(\log n)^2}\right). \tag{5.51}$$

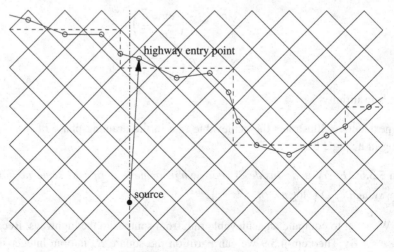

Fig. 5.12 Draining phase.

Now we note that as there are possibly many nodes in the squares, they have to share this bandwidth. Using Lemma 5.3.4, we conclude that the transmission rate of each node in the draining phase of our protocol is at least $R(d)/\log m$, which concludes the proof. $\square$

The following lemma illustrates the achievable rate on the multi-hop routes along the highway.

Lemma 5.3.7 *The nodes along the highways can w.h.p. achieve a per-node rate of* $\Omega\left(1/\sqrt{n}\right)$.

Proof We divide horizontal and vertical information flow, adopting the following multi-hop routing policy: pairwise coding and decoding is performed along horizontal highways, until we reach the crossing with the target vertical highway. Then, the same is performed along the vertical highways until we reach the appropriate exit point for delivery.

We start by considering the horizontal traffic. Let a node be sitting on the ith horizontal highway and compute the traffic that goes through it. Notice that, at most, the node will relay all the traffic generated in the ith slice of width w.

According to Lemma 5.3.5, a node on a horizontal highway must relay traffic for at most $2w\sqrt{n}$ nodes. As the maximal distance between hops is constant, by applying Theorem 5.3.2 we conclude that an achievable rate along the highways is $\Omega(1/\sqrt{n})$, with high probability.

The problem of the vertical traffic is the dual of the previous one. We can use the same arguments to compute the *receiving rate* of the nodes. Since each node is the destination of exactly one source, the rate per node becomes the same as above. $\square$

The following Lemma illustrates the achievable rate in the receiving phase of the protocol, occurring in a single hop.

Lemma 5.3.8 *Every destination node can w.h.p. receive information from the highway at rate* $\Omega\left((\log n)^{-3} n^{-\sqrt{2}c\kappa\gamma}/2\right)$.

Proof The delivery phase consists of communicating from the highway system to the actual destination. We proceed exactly in the same way as in Lemma 5.3.6, but in the other direction, that is, horizontal delivery from the vertical highways.

We divide the network area into vertical slices of constant width, and define a mapping between slabs and vertical paths. We assume that communication occurs from an *exit point* located on the highway, which is defined as the node of the vertical path closest to the horizontal line drawn from the destination. Again, the distance between exit points and destination is at most $\kappa \log m + \sqrt{2}c$. We can thus let $d = \kappa \log m + \sqrt{2}c$ in Corollary 5.3.3, and conclude that each square can be served at rate $R(d) = \Omega\left(n^{-\sqrt{2}c\kappa\gamma/2}/(\log n)^2\right)$. As there are at most $\log m$ nodes in each square by Lemma 5.3.4, the rate per node is at least equal to $R(d)/\log m$. $\square$

We are now ready to provide a proof of Theorem 5.3.1.

Proof of Theorem 5.3.1 We observe by Lemmas 5.3.6, 5.3.7, and 5.3.8, that if

$$\frac{\sqrt{2}}{2}c\kappa\gamma < \frac{1}{2}, \tag{5.52}$$

then the overall per-node rate is limited by the highway phase only, and the proof follows immediately from Lemma 5.3.7. Hence, we have to make sure that we can satisfy (5.52). Recall that $p = 1 - e^{-c^2}$ and by Theorem 4.3.8 that c and κ are constrained to be such that

$$c^2 > \log 6 + \frac{2}{\kappa}. \tag{5.53}$$

From (5.52) and (5.53) it follows that we can choose $\kappa = 1/2\sqrt{2}c\gamma$ and $c > (2\sqrt{2}\gamma + \sqrt{8\gamma^2 + \log 6})$ to conclude the proof. $\square$

5.4 Multiple source–destination pairs; information-theoretic upper bounds

In the previous section we have computed a lower bound on the achievable information rate per source–destination pair in random networks. To do this, we have explicitly described an operation strategy of the nodes that achieves the $1/\sqrt{n}$ bound w.h.p. We are now interested in finding a corresponding upper bound. We notice that as in the case of Theorem 5.2.2, to find an upper bound we cannot assume any restriction on the kind of help the nodes can give to each other: any user can act as a relay for any communicating pair, we are not restricted to pairwise coding and decoding along multi-hop paths, and arbitrary joint cooperation is possible. In this case the upper bound is *information theoretic*, in the sense that it follows from physical limitations, independent of the network operation strategy.

We consider, as usual, a Poisson point process of unit density inside the box B_n, and we partition B_n into two equal parts $[-\sqrt{n}/2, 0] \times [0, \sqrt{n}]$ and $[0, \sqrt{n}/2] \times [0, \sqrt{n}]$. We assume, as in the previous section, that there is a uniform traffic pattern: users are paired independently and uniformly, so that there are $O(n)$ communication requests that need to cross the boundary of the partition. Being interested in an upper bound, we also make the following optimistic assumptions.

Users on one side of the partition can share information instantaneously, and also can distribute the power among themselves in order to establish communication in the most efficient way with the users on the other side, which in turn are able to distribute the received information instantaneously among themselves.

We then introduce additional 'dummy' nodes in the network that do not generate additional traffic, but can be used to relay communications. That is, for each existing node x_k placed at coordinate (a_k, b_k), we place a dummy node y_k at mirror coordinate $(-a_k, b_k)$. Notice that an upper bound on the rate that the original Poisson points can achieve simultaneously to their intended destinations, computed under the presence of these extra nodes, is also an upper bound for the case when the extra nodes are not present. After introducing the dummy users, there are exactly the same number of nodes on each side of the partition of the box B_n. Furthermore, on each side, nodes are distributed according to a Poisson process with density $\lambda = 2$.

The channel model across the partition is the vectorial version of the Gaussian channel in (5.21); that is, for any configuration of n nodes placed in each side of the partition of the box B_n, we have

$$y_j(t) = \sum_{i=1}^{n} \ell(x_i, y_j) x_i(t) + z_j(t), \; j = 1, \ldots, n, \tag{5.54}$$

where $x_i(t)$ is the signal transmitted by node x_i on one side of the partition, $y_j(t)$ is the signal received by node y_j on the other side of the partition, $\ell(x_i, y_j)$ is the signal attenuation function between x_i and y_j and $z_j(t)$ a realisation of the additive white Gaussian noise process.

The above model shows that each node y_j on one side of the partition receives the signal from all the n nodes x_i on the other side, weighted by the attenuation factor $\ell(x_i, y_j)$, and subject to added white Gaussian noise z_j. Furthermore, if each node x_i has a power constraint $x_i^2(t) < P$, we have a total power constraint $\sum_{i=1}^{n} x_i^2(t) \leq nP$. Notice that the latter sum is over the n parallel Gaussian channels defined by (5.54).

Finally, we consider two kinds of attenuation functions: exponential signal attenuation of the type $e^{-(\gamma/2)|x-y|}$, with $\gamma > 0$; and power law signal attenuation of exponent $\alpha > 2$, $|x - y|^{-\alpha/2}$. We start by considering the exponential attenuation case first, and then treat the power law attenuation case.

Before proceeding, we state the information-theoretic cut-set bound, similar to Theorem 5.2.3, that we shall use to derive the desired upper bounds on the per-node rate. The reader is referred to Appendix A.6 for the definition of the singular values of a matrix.

Theorem 5.4.1 (**Cut-set bound**) *Consider an arbitrary configuration of $2n$ nodes placed inside the box B_n, partition them into two sets S_1 and S_2, so that $S_1 \cap S_2 = \emptyset$ and $S_1 \cup S_2 = S$, $|S_1| = |S_2| = n$. The sum $\sum_{k=1}^{n} \sum_{i=1}^{n} R_{ki}$ of the rates from the nodes $x_k \in S_1$ to the nodes $y_i \in S_2$ is upper bounded by*

$$\sum_{k=1}^{n} \sum_{i=1}^{n} R_{ki} \leq \max_{P_k \geq 0, \sum_k P_k \leq nP} \sum_{k=1}^{n} \log \left(1 + \frac{P_k s_k^2}{N} \right), \tag{5.55}$$

where s_k is the kth largest singular value of the matrix $L = \{\ell(x_k, y_i)\}$, N is the noise power spectral density, and as usual we have assumed the bandwidth $W = 1$.

Note that the location of the nodes enters the formula via the singular values s_k. Equation (5.55) should be compared to the broadcast-cut bound (5.31) and with the Shannon–Hartley formula (5.22). The broadcast-cut formula bounds the total flow of information from a single node to all other nodes in the network; the Shannon–Hartley formula gives the capacity of a single one-to-one transmission. The more general cut-set formula (5.55) bounds the total flow of information from all nodes on one side, to all nodes on the other side of the cut. It is not difficult to see that it is possible to recover (5.31) and (5.22) from (5.55); see the exercises. Finally, notice that while the Shannon–Hartley formula is achievable, (5.31) and (5.55) in general give only upper bounds on the sum of the rates. A proof of Theorem 5.4.1 follows classic information-theoretic arguments and is a direct consequence of Theorem 15.10.1 in the book by Cover and Thomas (2006); see also Telatar (1999).

5.4.1 Exponential attenuation case

The stage is now set to derive an upper bound on the achievable information rate $R(n)$ per source–destination pair in our random network. We let $C_n = \sum_{k=1}^{n} \sum_{i=1}^{n} R_{ki}$ be the sum of the rates across the partition, and we require all nodes to achieve rate $R(n)$ to their intended destination simultaneously. We notice that the upper bound on the sum of the rates provided by Theorem 5.4.1 depends on the singular values of the matrix $L = \{\ell(x_k, y_i)\}$, and hence on the actual locations of the nodes. The strategy is to just evaluate this bound for any arbitrary configuration of nodes, using a simple linear algebra argument, and then exploit the geometric structure of the the the Poisson point process to determine its asymptotic a.s. behaviour. Finally, an a.s. bound on the maximum achievable rate $R(n)$ per source–destination pair is obtained using the uniform traffic assumption. The reader is referred to Appendix A.6, for the necessary algebraic background.

Let us start by computing the upper bound on C_n. In this first part of the argument, we consider n arbitrary points $x_1, \ldots, x_n$, together with n mirror nodes, as explained above. First, notice that since the squares of the singular values of L coincide with the eigenvalues λ_k of the matrix LL^* (see Appendix A.6), it follows from (5.55) that

$$C_n \leq \sum_{k=1}^{n} \log\left(1 + \frac{nPs_k^2}{N}\right)$$

$$= \sum_{k=1}^{n} \log\left(1 + \frac{nP\lambda_k}{N}\right)$$

$$= \log \det\left(I + \frac{nP}{N}LL^*\right). \tag{5.56}$$

Using Hadamard's inequality $\det A \leq \prod_{k=1}^{n} A_{kk}$, which is valid for any non-negative definite matrix, we then obtain the following upper bound on C_n,

$$C_n \leq \sum_{k=1}^{n} \log\left(1 + \frac{nP(LL^*)_{kk}}{N}\right). \tag{5.57}$$

Next, we bound the diagonal elements of LL^*,

$$(LL^*)_{kk} = \sum_{i=1}^{n} |L_{ki}|^2$$

$$= \sum_{i=1}^{n} \ell(x_k, y_i)^2$$

$$\leq n e^{-\gamma |\hat{x}_k|}, \tag{5.58}$$

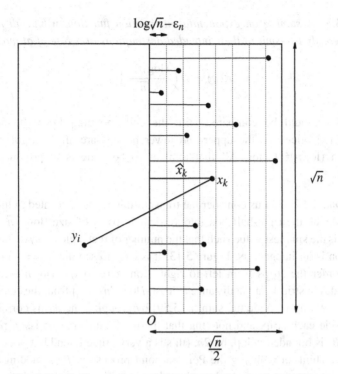

Fig. 5.13 Cutting the box in half and stripping the right-half box.

where $\widehat{x}_k$ is the first coordinate of x_k, see Figure 5.13, and where the inequality follows from the attenuation function being decreasing in the norm and by the triangle inequality. Substituting (5.58) into (5.57) we have

$$C_n \leq \sum_{k=1}^{n} \log\left(1 + \frac{n^2 P e^{-\gamma|\widehat{x}_k|}}{N}\right). \tag{5.59}$$

Notice that in (5.59) the bound on C_n explicitly depends on the geometric configuration of the points inside the box, i.e. on the distances of the points from the cut in the middle of the box. We now present the second part of the argument. Assuming the points to be distributed according to a Poisson point process, we can reasonably expect that most of them are far from this boundary, and hence the sum in (5.59) can be controlled. The following lemma makes this consideration precise.

Lemma 5.4.2 *Letting B_n' be a half-box of B_n, we have that w.h.p.*

$$C_n \leq \sum_{x_k \in X \cap B_n'} \log\left(1 + \frac{n^2 P e^{-\gamma|\widehat{x}_k|}}{N}\right) = O(\sqrt{n}(\log n)^2). \tag{5.60}$$

Clearly we also have that w.h.p.

$$R(n) = O\left(\frac{C_n}{n}\right). \tag{5.61}$$

Lemma 5.4.2, combined with (5.61), leads to the the following final result.

Theorem 5.4.3 *Assuming an exponential attenuation function, w.h.p. all nodes in B_n can simultaneously transmit to their intended destinations at a rate of at most*

$$R(n) = O\left(\frac{(\log n)^2}{\sqrt{n}}\right). \qquad (5.62)$$

Theorem 5.4.3 should be compared with the corresponding lower bound result of Theorem 5.3.1. Notice that the upper and lower bounds are almost tight, as they differ only by a $(\log n)^2$ factor. All that remains to be done is to provide a proof of Lemma 5.4.2.

Proof of Lemma 5.4.2 Let us consider the transmitting nodes x_k located in the right-half box. We subdivide the right-half box into vertical strips V_i of size $(\log \sqrt{n} - \epsilon_n) \times \sqrt{n}$, where $\epsilon_n > 0$ is the smallest value such that the number of rectangles $\sqrt{n}/(2(\log \sqrt{n} - \epsilon_n))$ in the partition is an integer; see Figure 5.13. It is easy to see that $\epsilon_n = o(1)$ as $n \to \infty$. Next, let us order the strips from left to right, from zero to $\sqrt{n}/(\log n - 2\epsilon_n) - 1$, and notice that nodes in strip V_i are at distance at least $|i \log \sqrt{n} - i\epsilon_n|$ from the cut in the centre of the box B_n. We now divide the sum in (5.59) by grouping the terms corresponding to the points inside each strip, and noticing that the total number of Poisson points inside the half box B'_n is bounded w.h.p. by $2n$. (this is a very crude bound but good enough for our purposes.) Think of realising the Poisson point process in B_n as adding a (random) number of points one by one. The bound we have for C_n increases with the number of points, and therefore we have w.h.p. that

$$C_n \leq \sum_{i=0}^{\frac{\sqrt{n}}{\log n - 2\epsilon_n}-1} \sum_{x \in X \cap V_i} \log\left(1 + \frac{(2n)^2 P e^{-\gamma|\bar{x}|}}{N}\right)$$

$$\leq \sum_{i=0}^{\frac{\sqrt{n}}{\log n - 2\epsilon_n}-1} \sum_{x \in X \cap V_i} \log\left(1 + \frac{P}{N} 4n^{2-\frac{1}{2}\gamma} e^{\gamma i \epsilon_n}\right), \qquad (5.63)$$

where the last inequality holds because nodes in strip V_i are at distance at least $|i \log \sqrt{n} - i\epsilon_n|$. We then split the sum into two parts and, letting $X(V_i)$ be the number of Poisson points falling inside strip V_i, we have for some constants K_1 and K_2

$$C_n \leq \sum_{i=0}^{\lceil 6/\gamma \rceil - 1} K_1 X(V_i) \log n + \sum_{i=\lceil 6/\gamma \rceil}^{\frac{\sqrt{n}}{\log n - 2\epsilon_n}-1} X(V_i) \log\left(1 + \frac{P}{N} 4n^{2-\frac{1}{2}\gamma} e^{\gamma i \epsilon_n}\right)$$

$$\leq \sum_{i=0}^{\lceil 6/\gamma \rceil - 1} K_1 X(V_i) \log n + \sum_{i=\lceil 6/\gamma \rceil}^{\frac{\sqrt{n}}{\log n - 2\epsilon_n}-1} K_2 \frac{X(V_i)}{n}. \qquad (5.64)$$

By applying Chernoff's bound, by the same computation as in Lemma 5.3.5, it is easy to see that w.h.p. we also have

$$X(V_i) \leq 2\sqrt{n} \log n, \text{ for all } i. \qquad (5.65)$$

It then follows, combining (5.64) and (5.65), that w.h.p.

$$C_n \leq O(\sqrt{n}\,(\log n)^2) + O\left(\frac{\sqrt{n}}{\log n - 2\epsilon_n}\,(\sqrt{n}\log n)\,\frac{1}{n}\right)$$
$$= O(\sqrt{n}\,(\log n)^2) + O(1), \qquad\qquad (5.66)$$

and the proof is complete. □

5.4.2 Power law attenuation case

A little more work is required to extend the result of Theorem 5.4.3 to a power law attenuation function. The main idea in this case is to adapt the stripping construction to the new shape of the attenuation function: previously, we have used strips of constant width when the attenuation was exponential; now, when the attenuation is a power law, to control the sum of the rates we need strips of width that decays exponentially with n. Once we define the new stripping of the box, the remaining part of the proof remains the same, although the computation is longer.

Theorem 5.4.4 *Assuming a power law attenuation function of exponent $\alpha > 2$, every node in B_n can w.h.p. (simultaneously) transmit to its intended destination at rate*

$$R(n) = O\left(\frac{n^{\frac{1}{\alpha}}(\log n)^2}{\sqrt{n}}\right). \qquad\qquad (5.67)$$

Notice that the bound provided in this case, for values of α close to two, is much weaker than the one in Theorem 5.4.3.

Proof of Theorem 5.4.4 As before, the proof is divided into two parts. First, we slightly modify the linear algebra argument to obtain a modified version of inequality (5.59). In a second stage, we modify the stripping used in Lemma 5.4.2 to accommodate for the different shape of the attenuation function and we repeat the computation of the sum, strip by strip. Let us start with the algebraic argument.

Thanks to the mirroring trick, the matrix L is symmetric. A standard, but tedious, computation shows that it can also be expressed as the convex combination of products of non-negative matrices, and hence is itself non-negative. Appendix A of Lévêque and Telatar (2005) shows this latter computation, which is not repeated here. A symmetric, non-negative definite matrix has non-negative, real eigenvalues which coincide with its singular values. Accordingly, we let the eigenvalues of L be $\mu_k \geq 0$, and $\mu_k = s_k$ for all k. We can then write a modified version of (5.56) as follows,

$$C_n \leq \sum_{k=1}^{n} \log\left(1 + \frac{nP\mu_k^2}{N}\right)$$
$$\leq \sum_{k=1}^{n} \log\left(1 + \sqrt{\frac{nP}{N}}\mu_k\right)^2$$

$$= 2 \sum_{k=1}^{n} \log \left(1 + \sqrt{\frac{nP}{N}} \mu_k \right)$$

$$= 2 \log \det \left(I + \sqrt{\frac{nP}{N}} L \right). \tag{5.68}$$

Again by Hadamard's inequality it now follows that

$$C_n \leq 2 \sum_{k=1}^{n} \log \left(1 + \sqrt{\frac{nP}{N}} L_{kk} \right)$$

$$= 2 \sum_{k=1}^{n} \log \left(1 + \sqrt{\frac{nP}{N}} \ell(x_k, y_k) \right),$$

$$= 2 \sum_{k=1}^{n} \log \left(1 + \sqrt{\frac{nP}{N}} (2|\widehat{x}_k|)^{-\frac{\alpha}{2}} \right). \tag{5.69}$$

We are now ready to show the second part of the proof which gives an a.s. upper bound on (5.69), assuming the nodes x_k to be Poisson distributed in the right-half box of B_n.

We subdivide the right-half box into vertical strips V_i, for $i = 1$ to $\lfloor \log(\sqrt{n}/2) + 1 \rfloor$, by drawing vertical lines at distance $\sqrt{n}/(2e^i)$ from the origin, for $i = 1, \ldots, \lfloor \log(\sqrt{n}/2) \rfloor$; see Figure 5.14. We want to compute the sum in (5.69) by grouping the terms

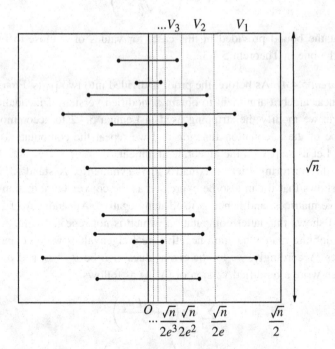

Fig. 5.14 Exponential stripping of the right-half box.

corresponding to the Poisson nodes inside each strip. Let us start considering the strip closest to the origin, $V_{\lfloor \log(\sqrt{n}/2)\rfloor+1}$. The contribution of the points in this strip to the sum is of $O(\sqrt{n}\log n)$, because there are w.h.p. $O(\sqrt{n})$ points inside this strip, and the minimum distance to the cut in the centre is w.h.p. at least $1/n^\delta$, for a fixed $\delta > 1/2$. This latter statement perhaps merits a little reflection. Consider a rectangle of width $n^{-\delta}$ and length $\sqrt{n}$ with its longest side coincident with the cut in the centre of the box. The probability that there are no points inside this rectangle is given by $\exp(-n^{1/2-\delta})$, since this number is Poisson distributed with parameter the area of the strip. Taking $\delta > 1/2$ the rectangle is empty w.h.p. and our claim on the minimum distance of the Poisson points from the cut holds.

We now want to bound the sum of the contributions of the points in all the remaining strips. To do this, we make the following three observations.

First, the points inside strip V_i, for $i = 1, \ldots, \lfloor \log(\sqrt{n}/2)\rfloor$ are at distance at least $\sqrt{n}/(2e^i) > 1$ from the centre axis of B_n. Second, there are w.h.p. at most $2n$ points in the right-half box of B_n. Third, by applying Chernoff's and the union bound it is also easy to see that w.h.p. we have

$$X(V_i) \le \frac{n}{e^i}\,(1 - 1/e) \text{ for all } i. \tag{5.70}$$

Reasoning as before, it now follows that the following bounds on (5.69) hold w.h.p.,

$$C_n \le 2 \sum_{i=1}^{\lfloor \log(\sqrt{n}/2)\rfloor} \sum_{x \in X \cap V_i} \log\left(1 + \sqrt{\frac{2nP}{N}(2|\widehat{x}|)^{-\frac{\alpha}{2}}}\right) + O(\sqrt{n}\log n)$$

$$\le 2 \sum_{i=1}^{\lfloor \log(\sqrt{n}/2)\rfloor} \sum_{x \in X \cap V_i} \log\left(1 + \sqrt{\frac{nP}{N}\left(\frac{\sqrt{n}}{e^i}\right)^{-\frac{\alpha}{2}}}\right) + O(\sqrt{n}\log n)$$

$$= 2 \sum_{i=1}^{\lfloor \log(\sqrt{n}/2)\rfloor} X(V_i)\,\log\left(1 + \sqrt{\frac{nP}{N}\left(\frac{\sqrt{n}}{e^i}\right)^{-\frac{\alpha}{2}}}\right) + O(\sqrt{n}\log n)$$

$$\le \frac{e-1}{e}\,2n \sum_{i=1}^{\lfloor \log(\sqrt{n}/2)\rfloor} \frac{1}{e^i}\log\left(1 + \sqrt{\frac{nP}{N}\left(\frac{\sqrt{n}}{e^i}\right)^{-\frac{\alpha}{2}}}\right) + O(\sqrt{n}\log n), \tag{5.71}$$

where the last inequality follows from (5.70).

We now let $\alpha/2 = \gamma > 1$, $M = \sqrt{n}$, and we compute an upper bound on the following sum (for notational convenience in the sequel we drop the $\lfloor \cdot \rfloor$ notation):

$$M^2 \sum_{i=1}^{\log M} \frac{1}{e^i}\log\left(1 + \frac{e^{\gamma i}}{M^{\gamma-1}}\right) = S_1 + S_2, \tag{5.72}$$

where

$$S_1 = M^2 \sum_{i=1}^{\frac{\gamma-1}{\gamma} \log M} \frac{1}{e^i} \log\left(1 + \frac{e^{\gamma i}}{M^{\gamma-1}}\right), \tag{5.73}$$

$$S_2 = M^2 \sum_{i=\frac{\gamma-1}{\gamma} \log M+1}^{\log M} \frac{1}{e^i} \log\left(1 + \frac{e^{\gamma i}}{M^{\gamma-1}}\right). \tag{5.74}$$

An upper bound on S_2 is easily obtained by substituting the smallest and the largest indices of the sum in the first and second product terms of (5.74) respectively, obtaining

$$S_2 = O(M^{\frac{\gamma+1}{\gamma}} (\log M)^2). \tag{5.75}$$

Notice that in our case, $M = \sqrt{n}$, so we have

$$S_2 = O(\sqrt{n}\, n^{1/\alpha} (\log n)^2). \tag{5.76}$$

We now focus on the sum S_1. By the Taylor expansion of the logarithmic function, we have

$$S_1 = M^2 \sum_{i=1}^{\frac{\gamma-1}{\gamma} \log M} \frac{1}{e^i} \sum_{k=1}^{\infty} \frac{(-1)^{k+1}}{k} \frac{e^{\gamma k i}}{M^{(\gamma-1)k}}$$

$$= M^2 \sum_{k=1}^{\infty} \frac{(-1)^{k+1}}{k} \frac{1}{M^{(\gamma-1)k}} \sum_{i=1}^{\frac{\gamma-1}{\gamma} \log M} e^{(\gamma k-1)i}. \tag{5.77}$$

We can compute the second sum in (5.77):

$$\sum_{i=1}^{\frac{\gamma-1}{\gamma} \log M} e^{(\gamma k-1)i} = e^{(\gamma k-1)} \frac{e^{(\gamma k-1)\left(\frac{\gamma-1}{\gamma} \log M\right)} - 1}{e^{(\gamma k-1)} - 1}$$

$$= \left(1 + \frac{1}{e^{(\gamma k-1)} - 1}\right)\left(M^{\frac{\gamma-1}{\gamma}(\gamma k-1)} - 1\right). \tag{5.78}$$

Substituting (5.78) into (5.77), we can split S_1 into $S_{11} + S_{12}$, where

$$S_{11} = M^2 \sum_{k=1}^{\infty} \frac{(-1)^{k+1}}{k} \frac{1}{M^{(\gamma-1)k}} \left(M^{\frac{\gamma-1}{\gamma}(\gamma k-1)} - 1\right)$$

$$= M^2 \frac{1}{M^{\frac{\gamma-1}{\gamma}}} \sum_{k=1}^{\infty} \frac{(-1)^{k+1}}{k} - M^2 \sum_{k=1}^{\infty} \frac{(-1)^{k+1}}{k} \left(\frac{1}{M^{\gamma-1}}\right)^k$$

$$= M^{\frac{\gamma+1}{\gamma}} \log 2 - M^2 \log\left(1 + \frac{1}{M^{\gamma-1}}\right)$$

$$= O\left(M^{\frac{\gamma+1}{\gamma}}\right) + O\left(M^{3-\gamma}\right)$$

$$= O\left(M^{\frac{\gamma+1}{\gamma}}\right), \tag{5.79}$$

where the last equality follows from $\gamma > 1$. The remaining part of the sum is given by

$$S_{12} = M^2 \sum_{k=1}^{\infty} \frac{(-1)^{k+1}}{k} \frac{1}{M^{(\gamma-1)k}} \left(M^{\frac{\gamma-1}{\gamma}(\gamma k-1)} - 1 \right) \frac{1}{e^{(\gamma k-1)} - 1}$$

$$\leq M^2 \sum_{k=1}^{\infty} \frac{1}{k} \frac{1}{M^{(\gamma-1)k}} \left(M^{\frac{\gamma-1}{\gamma}(\gamma k-1)} - 1 \right) \frac{1}{e^{(\gamma k-1)} - 1}$$

$$\leq M^2 \sum_{k=1}^{\infty} \frac{1}{k} \frac{1}{M^{(\gamma-1)k}} \left(M^{\frac{\gamma-1}{\gamma}(\gamma k-1)} - 1 \right) K e^{-(\gamma k-1)}$$

$$= M^{\frac{\gamma+1}{\gamma}} Ke \sum_{k=1}^{\infty} \frac{1}{k} (e^{-\gamma})^k - M^2 Ke \sum_{k=1}^{\infty} \frac{1}{k} \left(\frac{e^{-\gamma}}{M^{\gamma-1}} \right)^k$$

$$= M^{\frac{\gamma+1}{\gamma}} K \frac{e}{2} \log \left(\frac{1}{1-e^{-\gamma}} \right) + M^2 K \frac{e}{2} \log \left(1 - \frac{e^{-\gamma}}{M^{(\gamma-1)}} \right)$$

$$= O\left(M^{\frac{\gamma+1}{\gamma}} \right) + O\left(M^{3-\gamma} \right)$$

$$= O\left(M^{\frac{\gamma+1}{\gamma}} \right), \tag{5.80}$$

where the second inequality follows by choosing a sufficiently large constant K. We now note that when $M = \sqrt{n}$, by (5.79) and (5.80) we have

$$S_1 = O\left(\sqrt{n} \, n^{\frac{1}{\alpha}} \right). \tag{5.81}$$

Finally, by combining (5.76) and (5.81), we have

$$S_1 + S_2 = O(\sqrt{n} \, n^{1/\alpha} (\log n)^2). \tag{5.82}$$

The result now follows. $\qquad\square$

5.5 Historical notes and further reading

Information theory is an applied mathematics field started by Shannon (1948). The Shannon–Hartley capacity formula that we have presented here is perhaps the best known result in information theory, but it is only a special case of application of Shannon's general theory to a specific channel. For a more complete view see the books by Cover and Thomas (2006) and McEliece (2004). A general capacity cut-set bound can be found in Cover and Thomas (2006), see Theorem 14.10.1, and its application to the parallel Gaussian channel appears in Telatar (1999). Capacity scaling limits for single source–destination pairs presented here appear in Dousse, Franceschetti and Thiran (2006). The multiple source–destination pairs lower bound on the capacity is by Franceschetti, Dousse *et al.* (2007), while the computation of the upper bound is a variation of the approach of Lévêque and Telatar (2005), which appears in Franceschetti (2007). Capacity scaling limits of networks were first studied by Gupta and Kumar (2000), a work that sparked much of the interest in the field. Their original bounds gave the correct indication, but were derived in a slightly more restrictive non-information-theoretic setting. Xie and Kumar (2004) gave the first information-theoretic upper bounds and the proof of Lévêque and Telatar (2005) that we have presented here in slightly revised form, is a refinement of this work, although the obtained bound is not tight. The $1/\sqrt{n}$ lower bound of Franceschetti, Dousse *et al.* (2007) matches the upper bound of Xie and Kumar (2004) when the attenuation power law exponent is $\alpha > 6$, or the attenuation is exponential. Hence, there is no gap between capacity upper and lower bounds, at least up to scaling, in the high attenuation regime. Xue *et al.* (2005) and Özgür *et al.* (2007) studied the effect of non-deterministic loss functions on the per-node rate, and have shown that the inverse $\sqrt{n}$ law continues to hold in their models under the assumption of high attenuation over distance, while for low attenuation the added randomness permits a constant per-node rate.

Exercises

5.1 Give a formal proof of (5.23).

5.2 Provide a proof of Theorem 5.3.1 assuming an attenuation function of the type $\min\{1, x^{-\alpha}\}$ with $\alpha > 2$.

5.3 Check that $\lim_{R \to 0} \alpha(R) = 0$ in the proof of Theorem 5.2.2.

5.4 Derive the broadcast-cut bound (5.31) on the rates starting from the general cut-set bound (5.55). Hint: look at Equation (5.58).

5.5 Perform the computation leading to Equations (5.65) and (5.70).

5.6 The lower and upper bounds on the achievable rate in the multiple source–destination pairs for the high attenuation regime case differ by a factor of $(\log n)^2$, can you point out exactly where this factor arises in the computation of the upper bound?

5.7 In the upper bound given in Theorem 5.4.4, there is a $(\log n)^2$ term that arises from the computation of the sum S_2 in (5.74). Can you find a tighter bound on this sum removing a $\log n$ factor? Hint: this is tricky, your have to divide S_2 into multiple parts and develop an argument similar to the one used for S_1.

5.8 In Section 5.1.3 we have stated that if we assume ergodic inputs, then the codeword constraint (5.5) becomes the same as the practical constraint (5.17). Which theorem justifies this? We did not assume ergodic inputs to mathematically define the communication problem at the beginning of the chapter; nevertheless ergodicity is also used in the proof sketch of Theorem 5.1.4. Can you say why it is reasonable to use it in that context?

6

Navigation in random networks

In this chapter we shift our attention from the *existence* of certain structures in random networks, to the ability of *finding* such structures. More precisely, we consider the problem of navigating towards a destination, using only local knowledge of the network at each node. This question has practical relevance in a number of different settings, ranging from decentralised routing in communication networks, to information retrieval in large databases, file sharing in peer-to-peer networks, and the modelling of the interaction of people in society.

The basic consideration is that there is a fundamental difference between the *existence* of network paths, and their *algorithmic* discovery. It is quite possible, for example, that paths of a certain length exist, but that they are extremely difficult, or even impossible to find without global knowledge of the network topology. It turns out that the structure of the random network plays an important role here, as there are some classes of random graphs that facilitate the algorithmic discovery of paths, while for some other classes this becomes very difficult.

6.1 Highway discovery

To illustrate the general motivation for the topics treated in this chapter, let us start with some practical considerations. We turn back to the routing protocol described in Chapter 5 to achieve the optimal scaling of the information flow in a random network. Recall from Section 5.3 that the protocol is based on a multi-hop strategy along percolation paths that arise w.h.p. inside rectangles of size $m \times \kappa \log m$ that partition the entire network area. We have shown that if the model is highly supercritical, then for any κ, there are, for some $\delta > 0$, at least $\delta \log m$ disjoint crossing paths w.h.p. between the two shortest sides of each rectangle of the partition. We now make an important practical observation regarding these paths. In order to route information, each node must be able to decide which is its next-hop neighbour along the path. It is quite possible that if a node does not have a complete picture of the network topology inside a rectangle, it might route information in a 'wrong direction' that does not follow the proper crossing path. In other words, percolation theory ensures the existence of many disjoint crossing paths inside each rectangle, but in order to exploit them, nodes must 'see' these paths to perform the correct next-hop routing decision. It is interesting to ask whether it is still possible to route along the percolation paths without such global vision. Suppose, for example, that each node inside a rectangle only 'knows' the positions of the nodes falling inside a box of size $2\kappa \log m \times \kappa \log m$. Clearly, this is much less than seeing everything inside the whole rectangle, as in this case each node must know only the positions of roughly $\log^2 m$ other nodes rather than $m \log m$. We now ask the following question: is it possible, with only such limited knowledge, to route information along the paths crossing the whole rectangle? In the following, we answer this question in the affirmative, and point out the particular care that must be taken in describing the algorithmic procedure to navigate along crossing paths.

We start by showing the following corollary to Theorem 4.3.9. From now on we avoid the explicit indication of the ϵ_n when we consider integer partitions. This simplifies the notation and by now the reader should be able to fill in the formal details.

Corollary 6.1.1 *Consider bond percolation with parameter $p = 1 - e^{-c^2}$, as described in Section 5.3.1. Partition the network into small squares S_i of size $\kappa \log m \times \kappa \log m$. For all $\kappa > 0$ and c sufficiently large, there exist w.h.p. $(2\kappa/3) \log m$ disjoint open paths inside each square S_i that cross it from left to right.*

Proof This follows from (4.35), substituting $p = 1 - e^{-c^2}$ and $\delta = 2\kappa/3$. □

We now consider any three neighbouring squares S_l, S_m, and S_r, with S_l being the left-most of the three and S_r being the right-most of the three; see Figure 6.1. Since for all of them at least $(2\kappa/3) \log m$ nodes on the left side are connected to the right side via disjoint paths w.h.p., there are at least $(\kappa/3) \log m$ edge disjoint paths that cross from the left side of S_l to the right side of S_m and also $(\kappa/3) \log m$ edge disjoint paths that cross from the left side of S_m to the right side of S_r. Call these crossings *highway segments*. We can now use an 'interchange block' to connect the $(\kappa/3) \log m$ highway segments crossing a pair of adjacent squares with the next overlapping pair, as shown in Figure 6.1. A horizontal highway segment entering the middle of the three blocks in the figure will cut all vertical paths of the middle block. Order the highway segments

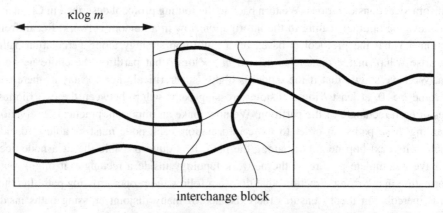

interchange block

Fig. 6.1   Since there are at least $(3\kappa/2) \log m$ paths crossing each box, there must be at least $(\kappa/3) \log m$ paths crossing any two adjacent boxes. These paths can be connected using an interchange.

entering the middle block from top to bottom, and consider now the ith one. The traffic on this highway segment can be routed onto the ith vertical path of the interchange, then onto the ith highway segment that exits the middle box from the right and crosses the whole next box. In this way, $(\kappa/3)\log m$ highways are constructed in each rectangle of size $m \times \kappa \log m$ by using one interchange in every square, and nodes in each square need only to know the positions of the nodes in two adjacent boxes.

Since the procedure for constructing the vertical highways proceeds along the same way, we conclude that it is possible to construct a complete highway system in the whole network, using only knowledge of topology over blocks of size of order $\log m \times \log m$ rather than $m \times \log m$. The main point to be taken from this reasoning is that to do so it is necessary to describe a specific *algorithmic* procedure, and this will be the main theme of this chapter.

6.2 Discrete short-range percolation (large worlds)

We now make a distinction between short-range percolation models and long-range ones. Short-range models exhibit geometric locality, in the sense that nodes are connected to close neighbours and no long-range connections exist. A typical example is the random grid, which is the model we focus on in this section. On the other hand, long-range percolation models add long-range random connections to an underlying subgraph which exhibits geometric locality. As we shall see, one peculiar difference between the two models is that the path length between nodes can be substantially smaller in long-range percolation models, which can form 'small world' networks, where almost all nodes are within a few hops of each other. On the other hand, small world networks can be difficult to navigate efficiently, as the algorithmic discovery of the short paths can be more difficult.

We start by introducing a random variable called the *chemical distance* between two points in a random network. Let us write $x \leftrightarrow y$ if there is a path connecting x to y.

Definition 6.2.1 *The chemical distance $C(x, y)$ between two nodes $x \leftrightarrow y$ in a random network is the (random) minimum number of edges forming a path linking x to y. If $x \nleftrightarrow y$ the chemical distance $C(x, y)$ is defined to be ∞.*

We now focus on the bond percolation model on the square lattice. The following result shows a large deviation estimate for nodes inside a connected component in the supercritical phase. Letting $|x - y|$ be the L_1 distance between x and y on $\mathbb{Z}^2$, and G the random grid, we have the following result.

Theorem 6.2.2 *For all $p > p_c$, and $x, y \in G$, there exist positive constants $c_1(p)$ and $c_2(p)$ such that for any $l > c_1|x - y|$, we have*

$$P(C(x, y) > l, x \leftrightarrow y) < e^{-c_2 l}. \tag{6.1}$$

Informally, this theorem says that in the supercritical phase, the chemical distance between two connected nodes is asymptotically of the same order as their distance on the fully connected lattice. In other words, the percolation paths that form above criticality behave almost as straight lines when viewed over larger scales. Notice that, as in the case of Theorems 4.3.3 and 4.3.4, what is remarkable here is that the statement holds for all $p > p_c$.

We do not give the proof of Theorem 6.2.2 here. Instead, we shift our attention from the existence of a path connecting two points, to the perhaps more practical question of *finding* such a path. We show that above criticality and for x, y with $|x - y| = n$, it is possible to find a path from x to y in order n steps in expectation. Notice that this implies a weaker version of Theorem 6.2.2, namely that the chemical distance is in expectation at most of order n.

Let us define a *decentralised algorithm* $\mathcal{A}$ as an algorithm that starts from node x and navigates towards a destination y, having only a limited view of the entire network. More precisely, initially $\mathcal{A}$ only knows the location of the destination y, the location of its starting point x, and of its immediate neighbours. At each step the algorithm 'hops' to one of its neighbours, and learns the location of the neighbours of this new position. We ask how many steps the algorithm must take before ever reaching the destination y. The following theorem shows that this number is on average of the same order as the chemical distance between source and destination.

Theorem 6.2.3 *For all $p > p_c$, and $x, y \in G$ such that $|x - y| = n$ and $x \leftrightarrow y$, there is a decentralised algorithm such that the expected number of steps required to reach y from x is $O(n)$.*

From the results above, we can conclude that above criticality it is possible to efficiently navigate the random grid, in the sense that provided that a path to the destination exists, its length grows at most linearly in the distance to the destination, and we can also find it on average in a linear number of steps. As we shall see shortly, this is not always the case, and in some models of random networks it can be difficult to find the shortest paths linking sources and destinations. Furthermore, we observe that our statement in Theorem 6.2.3 is weaker than what Theorem 6.2.2 says about the actual chemical distance, since we gave only a bound on the *expectation* of the number of steps performed by the algorithm.

Proof of Theorem 6.2.3 We explicitly construct an algorithm that achieves the desired bound. Let us fix one shortest path of length n connecting x and y on $\mathbb{Z}^2$. Notice that this choice can be made locally at each node according to some predefined deterministic rule. The algorithm tries to follow the shortest path until it encounters a closed edge. At this point the algorithm simply 'circumnavigates' the connected component in the dual graph that blocks the path, until either the destination point is reached, or the algorithm is back on the original shortest path to it; see Figure 6.2. Notice that this is always possible because x and y are assumed to be connected. The number of steps needed by the algorithm to reach the destination is bounded by the number of steps needed to circumnavigate at most n connected components in the dual graph. Recall from Exercise 4.1 that the average size of a connected component in the subcritical phase is an a.s. constant. It immediately follows that the number of steps in our algorithm is on average $O(n)$. $\qquad\square$

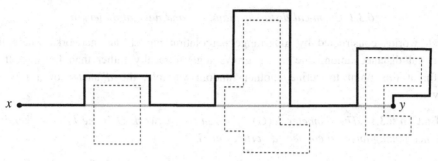

Fig. 6.2 Nodes x and y are connected following the shortest path (grey line) and circumnavigating the finite clusters of the dual graph (dashed line) that interrupt it.

6.3 Discrete long-range percolation (small worlds)

We now focus on long-range percolation models in which the chemical distance between any pair of nodes is much shorter than what Theorem 6.2.2 predicts. It is quite reasonable that by adding a few long-range connections, these can be used as 'shortcuts', and substantially decrease the chemical distance. This, as we shall see, does not mean in general that the short paths can be efficiently discovered by a decentralised algorithm.

The models we consider next add random connections to an underlying subgraph with probability decreasing in the distance between the nodes. If the parameters of this distribution are chosen in such a way that connections are sufficiently spread out, then in the resulting small world networks almost all nodes are within a few hops of each other. In this sense we say that these networks have a *weak geometric component*. As we discussed in Chapter 2, when connections are more spread out, we expect the model to gain some independence structure and to behave like an independent branching process. More precisely, we have shown that as nodes connect to neighbours that are further away, the percolation threshold of the model decreases, and in the limit it approaches the threshold of an independent branching process. Another consequence of spreading out the connections that we see here, is that the path length between nodes also decreases. If we are to compare a sufficiently spread-out long-range percolation process to a branching process that evolved for n steps, we see that in the latter case it is always possible to connect two nodes of the random tree by a path of length $\log n$, and it turns out that the same is true in our long-range percolation model.

Perhaps the extreme case of long-range percolation is when connections are added between all nodes at random, with probability independent of their distance lengths. This non-geometric model is a well studied one, originally considered by Erdös and Rényi in the late 1950s, and then extended by many authors since then, see for example the book by Bollobás (2001) for an extensive treatment. Not surprisingly, for a large class of probability distributions, this model also generates short paths among the nodes. What is perhaps more interesting about long-range percolation models is that our ability to find the short paths without any global knowledge of the network topology also changes depending on the probabilistic rules used to add the long-range connections.

6.3.1 Chemical distance, diameter, and navigation length

Small worlds constructed by long-range percolation models are networks where the chemical distance among the nodes grows logarithmically rather than linearly. It is useful at this point to define another random variable, the *diameter* of a random network,

Definition 6.3.1 *The diameter $D(G)$ of a random network G is the largest chemical distance among any two connected vertices in G.*

Notice that the diameter of a supercritical percolation model on the plane is infinite. Furthermore, for the random grid inside the box B_n of size $n \times n$ the diameter is w.h.p. at least as large as n, since this is the side length of the box, and we know by Theorem 4.3.4 that w.h.p. there are paths that cross B_n from side to side. We now show that the diameter can be drastically reduced if long-range connections are added to the full grid with a given probability distribution.

Let us consider a fully connected grid inside B_n and independently add long-range connections with probabilities that decay with the distance between the points. More precisely, we add a connection between grid points x and y with probability 1 if $|x - y| = 1$ and with probability $1 - \exp(-\beta/|x - y|^\alpha)$, if $|x - y| > 1$, where $\beta > 0$ and $\alpha > 0$ are fixed parameters. It is worth noticing that this latter connection function is close to one when x and y are close to each other, and decreases as a power law $\beta/|x - y|^\alpha$ when x and y are far away. Furthermore, notice that the number of edges incident to each node is a random variable which is not uniformly bounded. We show that the value of the diameter changes drastically when α changes from being larger than four to being smaller than four. The proof of the first result follows standard arguments, while the second result uses a self-similar construction that is reminiscent of the fractal percolation models as described, for example, in Meester and Roy (1996).

Theorem 6.3.2 *For $\alpha > 4$ there exists a constant $0 < \phi = \phi(\alpha) < \alpha - 4/\alpha - 3$ such that*

$$\lim_{n \to \infty} P(D(G_n) \geq n^\phi) = 1. \tag{6.2}$$

Proof The main idea of the proof is first to compute a bound on the length covered by the 'long' edges in the network, and then find a lower bound on the diameter by counting the number of hops required to cover the remaining distance to the destination using only the remaining 'short' edges.

Let $L(k)$ be the (random) total number of edges between pairs of points at distance k. We have that there exists a uniform constant C such that, for all n,

$$E(L(k)) \leq C \frac{n^2 k \beta}{k^\alpha}, \tag{6.3}$$

since the probability that any two nodes x and y at distance k are directly connected is $1 - \exp(-\beta/k^\alpha) \leq \beta/k^\alpha$. Fixing x, there are a constant times k nodes at distance k from x, and there are n^2 ways to choose x inside B_n.

Next, we compute the average sum over all points at distance $k > n^{1-\phi}$ of the number of edges between them weighted by their respective lengths,

$$
\begin{aligned}
E\left(\sum_{k>n^{1-\phi}} kL(k) \right) &= \sum_{k>n^{1-\phi}} kE(L(k)) \\
&\leq n^2\beta \sum_{k>n^{1-\phi}} k^{2-\alpha} \\
&= O(n^{2+(1-\phi)(2-\alpha+1)}),
\end{aligned} \tag{6.4}
$$

as $n \to \infty$, where the inequality follows from (6.3) and the last equality follows by substituting the lower value of $k = n^{1-\phi}$ inside the sum, considering an upper bound of n on the total number of terms in the sum, and letting $n \to \infty$. For the given value of $\phi < \alpha - 4/\alpha - 3$ we have that the exponent $2 + (1 - \phi)(3 - \alpha) < 1$ and hence we conclude that, as $n \to \infty$,

$$
E\left(\sum_{k>n^{1-\phi}} kL(k) \right) = o(n). \tag{6.5}
$$

By applying Markov's inequality we then have

$$
\lim_{n\to\infty} P\left(\sum_{k>n^{1-\phi}} kL(k) > n \right) = \lim_{n\to\infty} \frac{o(n)}{n} = 0. \tag{6.6}
$$

Notice that (6.6) bounds the total distance that shortcuts, i.e. long edges of length $k > n^{1-\phi}$, in the random grid can cover, to be at most n w.h.p. Since the largest distance between two points in B_n is $2n$, it then follows that w.h.p. any path between the two furthest points must use edges of length at most $n^{1-\phi}$ to cover the remaining distance n. It follows that any such path contains at least $n/n^{1-\phi} = n^\phi$ edges w.h.p., and the proof is complete. $\square$

Theorem 6.3.3 *For $\alpha < 4$ there exists a constant $\phi(\alpha) > 0$ such that*

$$
\lim_{n\to\infty} P(D(G_n) \leq (\log n)^\phi) = 1. \tag{6.7}
$$

Proof The proof is based on a 'self-similar' renormalisation argument. We partition B_n into subsquares s_i of side length n^γ with $\alpha/4 < \gamma < 1$. Let A_1 be the event that there exist at least two subsquares s_i and s_j such that there are no edges from s_i to s_j. For all i, let us now further subdivide each subsquare s_i into smaller squares s_{ik}, of side length n^{γ^2} and let A_2 be the event that there exists at least one s_i such that there are two subsquares of s_i which no not have an edge between them.

We iterate this m times in the natural way, obtaining in the end squares of side length n^{γ^m}. Assume now that none of the events $A_1, A_2, \ldots, A_m$ occurs. We claim that this implies that the diameter of the graph G_n is bounded by

$$
D(G_n) \leq 2^{m+2} n^{\gamma^m}. \tag{6.8}
$$

To show this claim, notice that since A_1 does not occur, we have $D(G_n) \leq 2 \max_i D(s_i) + 1$, because any two points in B_n are contained in at most two distinct subsquares and there is always one edge connecting the two subsquares. Similarly, since also A_2 does not occur, we have $D(G_n) \leq 4 \max_{i,k} D(s_{ik}) + 3$, see Figure 6.3. In the end, indicating by D_m the largest diameter of the subsquares of side length n^{γ^m}, we obtain that the diameter of our graph satisfies

$$
\begin{aligned}
D(G_n) &\leq 2^m D_m + 2^m - 1 \\
&\leq 2^{m+1} n^{\gamma^m} + 2^m \\
&\leq 2^{m+2} n^{\gamma^m},
\end{aligned}
\tag{6.9}
$$

where the second inequality follows from $D_m \leq 2n^{\gamma^m}$.

To complete the proof, we have to choose m such that the upper bound in (6.8) is at most $(\log n)^\phi$ and also w.h.p. none of the events $A_1, A_2, \ldots, A_m$ occurs. Let us consider the events A_i. We want to evaluate the probability that no edge exists between any two subsquares of side n^{γ^i} (we call these the *small subsquares*) that tessellate the square of side $n^{\gamma^{i-1}}$ (we call this the *large subsquare*). We note that the largest distance among any two points placed inside the large subsquare is bounded by $2n^{\gamma^{i-1}}$. Furthermore, there are a constant C times $n^{2\gamma^i}$ points in each small subsquare, forming at least $Cn^{4\gamma^i}$ pairs of possible connections between any two small squares. Since there are at most n^4 pairs of

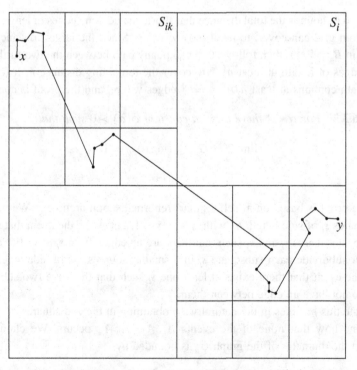

Fig. 6.3 If A_1 and A_2 do not occur, there is a path of length $4 \max D(s_{ik}) + 3$ that connects any two points x, y in the box B_n.

small subsquares inside the large subsquare, by the union bound it follows that as $n \to \infty$, the probability of A_i is bounded by

$$P(A_i) \leq n^4 \exp\left(-\frac{C\beta}{(2n^{\gamma^{i-1}})^\alpha}n^{4\gamma^i}\right)$$

$$= n^4 \exp\left(-Cn^{\gamma^{i-1}(4\gamma-\alpha)}\right). \qquad (6.10)$$

Furthermore, also by the union bound, (6.10), and the observation that $i-1 < m$ for all i (notice that $\gamma < 1$), we have

$$P(A_1 \cup A_2 \cup \cdots \cup A_m) \leq mn^4 \exp\left(-Cn^{\gamma^m(4\gamma-\alpha)}\right). \qquad (6.11)$$

We now want to choose m in such a way that as $n \to \infty$, (6.11) tends to zero and (6.8) is at most $(\log n)^\phi$. A possible choice to satisfy both of these conditions is

$$m = \frac{\log\log n - \log\log\log n + \log(4\gamma - \alpha) - \log K}{\log \gamma^{-1}} = O(\log\log n), \qquad (6.12)$$

where K is a large enough constant. Not believing in proof by intimidation, it is worth checking this latter statement. We have

$$\gamma^m = \gamma^{-\dfrac{\log\left(\frac{(4\gamma-\alpha)\log n}{K\log\log n}\right)}{\log\gamma}}$$

$$= \frac{K\log\log n}{(4\gamma - \alpha)\log n}, \qquad (6.13)$$

from which it follows that

$$\gamma^m(4\gamma - \alpha) = \frac{K\log\log n}{\log n}, \qquad (6.14)$$

$$n^{\gamma^m(4\gamma-\alpha)} = (\log n)^K, \qquad (6.15)$$

$$\exp\left(-Cn^{\gamma^m(4\gamma-\alpha)}\right) = n^{-CK}, \qquad (6.16)$$

and by substituting (6.16) into (6.11) we can then choose K large enough so that $P(A_1 \cup A_2 \cup \cdots \cup A_n)$ tends to zero. As for (6.8), using (6.15), we have that

$$D(G_n) \leq 2^{m+2}n^{\gamma^m}$$

$$= 2^{m+2}(\log n)^{\frac{K}{4\gamma-\alpha}}$$

$$= (\log n)^\phi. \qquad (6.17)$$

$\square$

We have shown in Theorem 6.3.3 that when $\alpha < 4$ the diameter of the long-range percolation model is w.h.p. at most $O((\log n)^\phi)$, for some $\phi > 0$. We now show that when $\alpha > 2$, a navigation algorithm that at each step has only limited knowledge of the network topology cannot find a path of this order of length. This means that for $2 < \alpha < 4$, there is a gap between the diameter of the long-range percolation model on the one hand, and what can be achieved by any decentralised algorithm on the other hand. Let us first define the *navigation length* of the graph as follows.

Definition 6.3.4 *Given a random network G and a navigation algorithm $\mathcal{A}$, the navigation length $D(\mathcal{A})$ is the (random) minimum number of steps required by $\mathcal{A}$ to connect two randomly chosen vertices on G.*

Notice that in the above definition we have considered the number of steps required to connect two *random* points. On the other hand, in Definition 6.3.1 we have considered the worst case scenario of the largest chemical distance between *any* two points. We have opted for these choices because while the diameter is a property of the network itself, the navigation length also depends on the algorithm, and picking two random points seemed more natural to give an illustration of the average performance of the algorithm. It is easy to check however, and it is a good exercise to do so, that the next theorem also holds if we employ the worst case scenario definition for the navigation length. Notice also that it is always the case that $D(\mathcal{A}) \geq D(G)$. The next theorem also illustrates one instance in which the strict inequality holds. Recall that for $\alpha < 4$, the diameter is w.h.p. $O\left((\log n)^\phi\right)$, for some $\phi > 0$.

Theorem 6.3.5 *For all $\alpha > 2$, $\phi < (\alpha - 2)/(\alpha - 1)$ and decentralised algorithm $\mathcal{A}$, we have*

$$\lim_{n \to \infty} P(D(\mathcal{A}) \geq n^\phi) = 1. \tag{6.18}$$

Proof Call the $n \times n$ grid G_n, and consider a node $x \in G_n$. For $r > 1$, the probability that x is directly connected to at least one node $y \in G_n$ with $|x - y| > r$ is bounded above by

$$\sum_{y \in G_n : |x-y| > r} \frac{\beta}{|x-y|^\alpha} \leq \sum_{k=r+1}^{\infty} \frac{\beta 4k}{k^\alpha}$$

$$\leq 4\beta \int_r^\infty x^{1-\alpha} dx$$

$$= \frac{4\beta}{\alpha - 2} r^{2-\alpha}. \tag{6.19}$$

Now pick two nodes uniformly at random on G_n and consider the path that algorithm $\mathcal{A}$ finds between them. Notice that w.h.p. the distance between the two randomly chosen nodes is at least $n^{1-\epsilon}$, for any $\epsilon > 0$. It follows that if the path contains at most n^ϕ steps, then w.h.p. there must be one step of length at least $n^{1-\epsilon}/n^\phi = n^{1-\epsilon-\phi}$. We now compute a bound on the probability of the event A_n that such a step appears in the first n^ϕ steps of the algorithm. By the union bound and (6.19), we have

$$P(A_n) \leq n^\phi \frac{4\beta}{\alpha - 2} n^{(1-\epsilon-\phi)2-\alpha}. \tag{6.20}$$

The exponent of the expression above can be made less than zero by choosing $\phi < (\alpha - 2)/(\alpha - 1)$ for sufficiently small ϵ. It immediately follows that $P(A_n) \to 0$ as $n \to \infty$

and hence a route with fewer than n^ϕ hops cannot be found w.h.p. and the proof is complete. ☐

6.3.2 More on navigation length

We now look at the navigation length of other long-range percolation models. We start with a discrete model first, similar to the one considered in the previous section, but in which the number of long-range connections of each node is a given constant. In the next section we consider some natural continuum versions of these models, which exhibit similar features. It turns out that all of these models show a threshold behaviour at $\alpha = 2$, and at the end of the chapter we provide an informal argument to explain this peculiar behaviour, introducing the notions of scale invariance and universality.

In the first model of this section, we add a constant number l of directed long-range random connections to each node x in the full grid inside B_n. These connections are directed edges between points of the grid, each one added independently between x and y with probability $|x - y|^{-\alpha}/(\sum_y |x - y|^{-\alpha})$, where the sum is over all grid points y inside B_n. The model has the same geometric interpretation as the previous one, in the sense that the long-range connections of each node are distributed broadly across the grid, with probabilities that decay as a power law of the distance to the destination. It is easy to see that when $\alpha = 0$ the long-range contacts are uniformly distributed, while as α increases, the long-range contacts of a node become more and more clustered in its vicinity on the grid. We have the following theorem.

Theorem 6.3.6 *For the discrete long-range percolation model described above, the following statements hold.*

(i) *For $\alpha = 2$ and $l = 1$, there exists a decentralised algorithm $\overline{A}$ and a constant K, such that $E(D(\overline{A})) \leq K(\log n)^2$. Furthermore, we also have that for any $\epsilon > 0$, there exists a $K' > 0$ such that*

$$\lim_{n \to \infty} P(D(\overline{A}) \leq K'(\log n)^{2+\epsilon}) = 1. \tag{6.21}$$

(ii) *For any $\alpha < 2$, $\phi(\alpha) < (2 - \alpha)/3$, $l \geq 0$, and decentralised algorithm A, we have*

$$\lim_{n \to \infty} P(D(A) > n^\phi) = 1. \tag{6.22}$$

(iii) *For any $\alpha > 2$, $\phi(\alpha) < (\alpha - 2)/(\alpha - 1)$, $l \geq 0$, and decentralised algorithm A, we have*

$$\lim_{n \to \infty} P(D(A) > n^\phi) = 1. \tag{6.23}$$

Proof   Case (i). We consider the following algorithm $\overline{A}$: at each step, node x holding the message sends it to the node as close to the target t as possible (in L_1 distance). We start by noticing the following deterministic bounds that hold on the $n \times n$ square grid G_n, for any node $x \in G_n$:

$$\sum_{y \in G_n, y \neq x} |x - y|^{-2} \leq \sum_{j=1}^{2n-2} 4j \, j^{-2}$$

$$= 4 \sum_{j=1}^{2n-2} j^{-1}$$

$$\leq 4 + 4 \log(2n - 2)$$

$$\leq 4 \log(6n). \tag{6.24}$$

Hence, we have a lower bound of

$$\frac{|x - y|^{-2}}{4 \log(6n)} \tag{6.25}$$

on the probability that node x chooses node y as its long-range contact, at any given step of the algorithm.

We now make the following remarks regarding our algorithm. First, since the distance to the target strictly decreases at each step, each node receives the message at most once, i.e., there are no loops in the path to the destination and this preserves independence in successive steps in the algorithm. Second, we say that the algorithm is in *phase* j if the distance from the current node to the target is greater than 2^j and at most 2^{j+1}. It is clear that the initial value of j is at most $\log n$ and that when $j < \log \log n$ the algorithm can deliver the message to the target in at most $\log n$ steps.

Let us now assume that $j \in [\log \log n, \log n]$, and node x has the message. We ask how many steps are required to complete this phase. We first compute a lower bound on the probability of the event A_j that phase j ends at the first step, i.e., the probability that node x sends the message into the set D_j of nodes that are within lattice distance 2^j of the target t. It is not hard to see that the number of nodes in D_j is bounded below by 2^{2j}. Each node in D_j is within lattice distance $2^{j+1} + 2^j < 2^{j+2}$ of x, see Figure 6.4, and if any one of these nodes is the (only) long-range contact of x, the message will be sent to the interior of D_j. By summing the individual probabilities and using (6.25), we have

$$P(A_j) \geq \frac{2^{2j}}{4 \log(6n) 2^{2j+4}} = \frac{1}{64 \log(6n)}. \tag{6.26}$$

If x does not have such a shortcut, the message is passed to a short-range contact which is closer to the target and the same lower bound on the probability of having a shortcut into D_j holds at the next step. Hence, the number of steps spent in phase j until a suitable long-range connection is found is upper bounded by a geometric random variable S_j with mean

$$\frac{1}{P(A_j)} = O(\log n). \tag{6.27}$$

It follows that phase j is completed on average in $O(\log n)$ steps. Since there are at most $\log n$ phases to complete, the total number of steps is on average at most $O((\log n)^2)$.

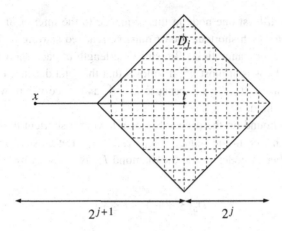

Fig. 6.4   Sketch of phase j.

We also want to bound the total number of steps w.h.p., so we let $\epsilon > 0$ and notice that

$$P(S_j \leq 64(\log 6n)^{1+\epsilon}) \geq 1 - \left(1 - \frac{1}{64 \log 6n}\right)^{64(\log 6n)^{1+\epsilon}}$$

$$= 1 - e^{-(\log 6n)^{\epsilon}}$$

$$= 1 - \left(\frac{1}{6n}\right)^{\epsilon} \to 1. \qquad (6.28)$$

It immediately follows that w.h.p. the number of steps to complete each phase is $O((\log n)^{1+\epsilon})$, and the total number of steps is $O((\log n)^{2+\epsilon})$, so the proof of this case is complete.

Case (ii). Let us select the source s and the target t uniformly at random on the grid. We start by noticing the following deterministic bound that holds for any node x on the square grid:

$$\sum_{y \in G_n, y \neq x} |x - y|^{-\alpha} \geq \sum_{j=1}^{n/2} j^{1-\alpha}$$

$$\geq \int_1^{n/2} x^{1-\alpha} dx$$

$$= \frac{(n/2)^{2-\alpha} - 1}{2 - \alpha}. \qquad (6.29)$$

We now let D_δ^n be the diamond centred at t and of radius n^δ, for some $\delta \in (\phi, 1)$. It is clear that for any $\epsilon > 0$, the distance from s to the diamond D_δ^n is larger than $n^{1-\epsilon}$ w.h.p. and therefore the source will be outside D_δ^n w.h.p.

We reason by contradiction and assume that there exists an algorithm which can route from s to t in fewer than n^ϕ hops. Accordingly, we let the sequence of nodes visited by the algorithm be $s = x_0, x_1, \ldots, x_m = t$, with $m \leq n^\phi$. We claim that w.h.p. there must be a shortcut from at least one node in this sequence to the interior of the diamond D_δ^n. Indeed, if there is no such shortcut, then t must be reached starting from a node outside D_δ^n and using only short-range links. But since the length of each short-range link is one and the number of hops is at most n^ϕ, it follows that the total distance travelled by using only local hops is at most $n^\phi < n^\delta$, because $\delta > \phi$, and since w.h.p. we started outside D_δ^n, our claim must hold.

Next, we find a bound on the probability of a having a shortcut to the interior of the diamond D_δ^n at each step in the sequence $x_0, x_1, \ldots, x_m$. Let us start by focusing on the point x_0. The number of nodes inside the diamond D_δ^n is bounded by

$$|D_\delta^n| \le 1 + \sum_{j=1}^{n^\delta} 4j \le 4n^{2\delta}, \qquad (6.30)$$

where we assume that n is large enough so that $n^\delta \ge 2$. Letting LR_δ be the event that x_0 has a long-range connection to at least one of the nodes in D_δ, by (6.29) we have

$$
\begin{aligned}
P(LR_\delta) &\le \frac{l|D_\delta^n|}{((n/2)^{2-\alpha} - 1)/(2 - \alpha)} \\
&\le \frac{l(2 - \alpha)4n^{2\delta}}{(n/2)^{2-\alpha} - 1} \\
&= O(n^{2\delta - 2 + \alpha}),
\end{aligned}
\qquad (6.31)
$$

where we have assumed $l > 0$, since for $l = 0$ the theorem clearly holds. If x_0 does not have such a shortcut, the message is passed to x_1 which is a short-range contact closer to the target and hence the upper bound on the probability of having a shortcut into D_δ^n holds in this case. By iteration, we have that the same upper bound holds at every step in the sequence. Now, by letting LR^ϕ be the event that LR_δ occurs within n^ϕ hops and applying the union bound, we have

$$P(LR^\phi) = O(n^{\phi + 2\delta - 2 + \alpha}), \qquad (6.32)$$

which tends to zero as $n \to \infty$, provided that $\phi < (2 - \alpha)/3$ and choosing $\delta > \phi$ small enough such that $\phi + 2\delta - 2 + \alpha < 0$. This leads to a contradiction and the proof of this case is complete.

Case (iii). The proof of this case is similar to the proof of Theorem 6.3.5. Consider a node $x \in G_n$ and let y be a randomly generated long-range contact of x. For any $m > 1$, we have

$$
\begin{aligned}
P(|x - y| > m) &\le \sum_{j=m+1}^{\infty} 4j \, j^{-\alpha} \\
&= 4 \sum_{j=m+1}^{\infty} j^{1-\alpha}
\end{aligned}
$$

$$\leq \int_m^\infty x^{1-\alpha} dx$$

$$= \frac{m^{2-\alpha}}{\alpha - 2}. \tag{6.33}$$

Now, pick two nodes at random and consider the path that algorithm $\mathcal{A}$ finds between them. Notice that w.h.p. the distance between the two randomly chosen nodes is at least $n^{1-\epsilon}$ for any $\epsilon > 0$. It follows that if the path contains at most n^ϕ steps, then there must be one step of length at least $m = n^{1-\epsilon}/n^\phi = n^{1-\epsilon-\phi}$. By the union bound and (6.33), the probability of the event A_n that such a step appears in the first n^ϕ steps of the algorithm, is bounded by

$$P(A_n) \leq l \frac{n^\phi n^{(1-\epsilon-\phi)^{2-\alpha}}}{\alpha - 2}. \tag{6.34}$$

When $\phi < (\alpha - 2)/(\alpha - 1)$, the exponent of the expression above can be made less than zero by choosing ϵ sufficiently small. It immediately follows that in that case, $P(A_n) \to 0$ as $n \to \infty$ and hence a route with fewer than n^ϕ hops cannot be found w.h.p. and the proof is complete. $\square$

6.4 Continuum long-range percolation (small worlds)

We now consider models which are defined on the continuum plane. The first one adds random long-range connections to a fully connected boolean model inside the box B_n of side length $\sqrt{n}$. For ease of exposition, here we consider distances as being defined on the torus obtained by identifying opposite edges of B_n. This means that we do not have to deal with special cases occurring near the boundary of the box, and that events inside B_n do not depend on the particular location inside the box.

Let X be a Poisson point process of unit density inside the box and let the radius r of the boolean model be $\sqrt{c \log n}$, where c is a sufficiently large constant so that the model is fully connected w.h.p. Notice that this is possible by Theorem 3.3.4. This boolean model represents the underlying short-range connected graph. We then add undirected long-range connections between points $x, y \in X$ such that $r(x, y) = |x - y| > \sqrt{c \log n}$, in such a way that an edge is present between nodes x and y with probability $\min\{\beta_n r(x, y)^{-\alpha}, 1\}$, where β_n determines the expected node degree l.

Notice that the continuum model described above is conceptually similar to the discrete one in the previous section. However, there are some important differences that should be highlighted. In the first place, the underlying short-range graph is not deterministic as in the previous case, but random. Furthermore, the number of long-range connections is also random in this case. These differences lead to some dependencies that must be carefully dealt with, and the analysis becomes considerably more complicated. Nevertheless, it turns out that a similar result as in Theorem 6.3.6 holds also in this case.

Theorem 6.4.1 *For the continuum long-range percolation model described above, the following statements hold.*

(i)   *For $\alpha = 2$, $l = 1$, and sufficiently large c, there exists a decentralised algorithm $\overline{\mathcal{A}}$ and a constant K, such that $E(D(\overline{\mathcal{A}})) \leq K(\log n)^2$. Furthermore, we also have that for any $\epsilon > 0$, there exists a $K' > 0$ such that*

$$\lim_{n \to \infty} P(D(\overline{\mathcal{A}}) \leq K'(\log n)^{2+\epsilon}) = 1. \qquad (6.35)$$

(ii) *For all $\alpha < 2$, $\phi(\alpha) < (2 - \alpha)/6$, decentralised algorithm $\mathcal{A}$, and sufficiently large l and c, we have*

$$\lim_{n \to \infty} P(D(\mathcal{A}) > n^{\phi}) = 1. \qquad (6.36)$$

(iii) *For all $\alpha > 2$, $\phi(\alpha) < (\alpha - 2)/(2(\alpha - 1))$, decentralised algorithm $\mathcal{A}$, and sufficiently large l and c, we have*

$$\lim_{n \to \infty} P(D(\mathcal{A}) > n^{\phi}) = 1. \qquad (6.37)$$

Proof Case (i). We consider the following algorithm: at each step, if node x holding the message can find a long-range connection which reduces the distance to the target by a factor of at least $1/2$, but no larger than $3/4$, then it sends the message along such connection. If there are several such long-range connections, then one of them is chosen at random. If there is none, then the algorithm uses a short-range connection that reduces the distance to the destination.

We make the following two observations concerning this algorithm. First, we notice that to ensure that our algorithm does not get stuck, it requires that node x is able to find, at each step, a short-range contact closer to the final target point t than itself, and we will show in a moment that this is true w.h.p. by choosing the constant c of the model large enough. Second, we notice that the reason we avoid using long-range connections that reduce the distance to the target by a factor larger than $3/4$, is to preserve independence in the analysis of successive steps of the algorithm. If x were simply to route the message to the node y that is the closest to the target among its neighbours, then in the analysis of the next step, the conditional law of the point process in the circle centred at the target and of radius $|t - y|$ would no longer be Poisson. The fact that we know there are no connections from x to this circle biases the probability law. On the other hand, if at one step of the algorithm we proceed without looking for connections that reduce the distance to the target by a factor larger than $3/4$, then at the next step we do not know anything about the connections of the points inside the disc of radius $r/4$ (r is the distance between source and target) centred at the target, and hence we can safely repeat the analysis starting from the new point y.

In the following, we denote by $C(u, r)$ the disc of radius r centred at node u, and by $A(t, r)$ the annulus $C(t, r/2) \backslash C(t, r/4)$. Initially r has value equal to the distance between s and t.

We now prove that the algorithm does not get stuck. Consider the discs $C_1 = C(x, \sqrt{c \log n})$ and $C_2 = C(t, |x - t|)$. For any point $y \in C_1 \cap C_2$ we have that $|y - t| < |x - t|$. Moreover, the intersection contains a sector of C_1 of angle at least $2\pi/3$ that we denote by D; see Figure 6.5.

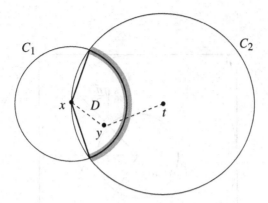

Fig. 6.5 The sector D is of angle at least $2\pi/3$, since the radius of C_1 is smaller than the radius of C_2.

Now partition B_n into smaller subsquares s_i of side length $a\sqrt{c \log n}$ and notice that we can choose the constant a, independent of c, such that D fully contains at least one of the small subsquares. It follows that if every s_i contains at least one point of the point process, then every node at distance greater than $\sqrt{c \log n}$ from t can find at least one short-range contact which is closer to t than itself. We call $X(s_i)$ the number of Poisson points inside subsquare s_i. By the union bound, we have

$$P(X(s_i) \geq 1, \text{ for all } i) \geq 1 - \sum_{i=1}^{n/(a^2 c \log n)} P(X(s_i) = 0)$$

$$= 1 - \frac{n}{a^2 c \log n} e^{-a^2 c \log n}$$

$$\to 1, \tag{6.38}$$

as $n \to \infty$, by choosing c large enough that $a^2 c > 1$.

Now that we know that our algorithm does not get stuck w.h.p., we proceed to show that it reaches the target in $O\left((\log n)^2\right)$ steps w.h.p. Let us first find a bound on the normalisation factor β_n by computing the following bound on the expected number of long-range connections l. With reference to Figure 6.6, we have

$$l \leq \beta_n \int_{\sqrt{c \log n}}^{\sqrt{n/2}} x^{-2} 2\pi x \, dx = \pi \beta_n (\log n - \log \log n - \log(2c)). \tag{6.39}$$

Since $l = 1$, it follows that

$$\beta_n \geq \frac{1}{\log n}. \tag{6.40}$$

We now compute the probability of finding a suitable shortcut at a generic step of the algorithm. Let $r > \sqrt{c \log n}$ and let N_A be the number of nodes in the annulus $A(t, r)$.

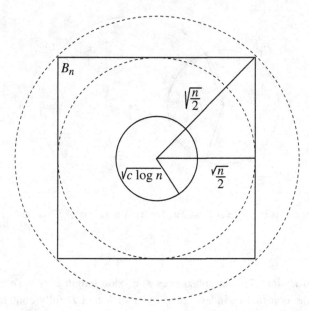

Fig. 6.6 Computation of the average number of long-range connections per node.

Notice that this is a Poisson random variable with parameter $3\pi r^2/16$. By Chernoff's bound in Appendix A.4.3, it is easy to see that

$$P\left(N_A \le \frac{3\pi r^2}{32}\right) \le \exp\left(-\frac{3\pi c \log n}{32}(1-\log 2)\right). \tag{6.41}$$

Furthermore, since the distance from x to any node in $A(t, r)$ is at most $3r/2$, letting LR be the event that x has a long-range connection to at least one of the N_A nodes in $A(t, r)$, we have

$$P(LR|N_A = k) \ge 1 - \left(1 - \frac{4\beta_n}{9r^2}\right)^k. \tag{6.42}$$

We denote the number at the right-hand side by γ_k^n. Observe that the bound really is a worst case scenario in the following sense: if we condition on any event E contained in $\{N_A \ge M\}$, then the conditional probability of LR given E will be at least $\gamma_M^n = 1 - (1 - 4\beta_n/(9r^2))^M$, whatever other information E contains. Indeed, if E is contained in $\{N_A \ge M\}$, then the 'worst' that can happen for LR is that there are exactly M points, and that all these points are maximally far away. This leads to $P(LR|E) \ge \gamma_M^n$.

If x does not have a shortcut to $A(t, r)$, the message is passed to a short-range contact which is closer to the target. At this moment we condition on $N_A = k$, *and* on the fact that x did not have a shortcut. Since this event is contained in the event that $N_A = k$, the (conditional) probability that the next point does have an appropriate shortcut satisfies the same lower bound γ_k^n as before, according to the observation made above.

Iterating this, we see that conditioned on $N_A = k$, the number S_x of short-range hops until a suitable shortcut is found, is at most a geometric random variable with mean $1/\gamma_k^n$, and therefore

$$E(S_x|N_A = k) \leq \frac{1}{\gamma_k^n}. \tag{6.43}$$

We now write

$$E(S_x) = \sum_{k \leq \frac{3}{32}\pi r^2} E(S_x|N_A = k)P(N_A = k)$$

$$+ \sum_{k > \frac{3}{32}\pi r^2} E(S_x|N_A = k)P(N_A = k)$$

$$= W_1 + W_2. \tag{6.44}$$

The first sum is bounded as

$$W_1 \leq n \sum_{k \leq \frac{3}{32}\pi r^2} P(N_A = k)$$

$$\leq n \exp\left(-\frac{3\pi c \log n}{32}(1 - \log 2)\right), \tag{6.45}$$

where the first inequality holds because, conditioned on $N_A = k$, the average number of points outside the annulus $A(t, r)$ is at most n, and the second inequality follows from (6.41). Notice that by choosing c large enough, (6.45) tends to zero as $n \to \infty$. We now want to bound the sum W_2. It follows from (6.43) and (6.42) that

$$W_2 \leq \sum_{k > \frac{3}{32}\pi r^2} \frac{P(N_A = k)}{\gamma_{\frac{3\pi r^2}{32}}^n}$$

$$\leq \frac{1}{\gamma_{\frac{3\pi r^2}{32}}^n}. \tag{6.46}$$

We now notice that by (6.40) and using the inequality $(1 - x)^n \leq (1 - nx/2)$, which holds for x sufficiently small, we have

$$\gamma_{\frac{3\pi r^2}{32}}^n = 1 - \left(1 - \frac{4\beta_n}{9r^2}\right)^{\frac{3\pi r^2}{32}}$$

$$\geq \frac{4}{9r^2 \log n} \frac{3\pi r^2}{32} \frac{1}{2}$$

$$= \frac{\pi}{48 \log n}. \tag{6.47}$$

Finally, combining things together yields

$$E(S_x) \leq n \exp\left(-\frac{3\pi c \log n}{32}(1 - \log 2)\right) + \frac{48 \log n}{\pi}$$

$$= o(1) + O(\log n), \tag{6.48}$$

where the last equality follows by choosing c large enough.

Finally, we notice that the total number of shortcuts needed to reach the target is at most of order $\log n$, since the initial value of r is at most $\sqrt{2n}$ and r decreases by a factor of at least $1/2$ each time a shortcut is found. It immediately follows that the expected total number of hops until $r < \sqrt{c \log n}$ is of order $(\log n)^2$.

The second claim follows from a straightforward application of Chebyshev's inequality; see the exercises.

Case (ii). We start by computing a bound on the normalisation factor β_n. With reference to Figure 6.6, we have

$$l \geq \beta_n \int_{\sqrt{c \log n}}^{\frac{\sqrt{n}}{2}} x^{-\alpha} 2\pi x \, dx = \frac{2\pi \beta_n}{2 - \alpha} \left(\frac{n^{(2-\alpha)/2}}{2^{2-\alpha}} - (c \log n)^{(2-\alpha)/2} \right), \tag{6.49}$$

from which it follows that

$$\beta_n \leq \frac{4l}{n^{(2-\alpha)/2}}. \tag{6.50}$$

Let us select the source s and destination t uniformly at random and let $C_\delta^n = C(t, n^\delta)$ for some $\delta \in (\phi, 1/2)$. For any $\epsilon > 0$, the distance from s to the disc C_δ^n is larger than $n^{(1/2)-\epsilon}$ w.h.p. We reason by contradiction and assume that there exists an algorithm which can route from s to t in fewer than n^ϕ hops. Accordingly, we let the sequence of nodes visited by the algorithm be $s = x_0, x_1, \ldots, x_m = t$, with $m \leq n^\phi$. We claim that there must be a shortcut from at least one node in this sequence to the interior of the circle C_δ^n. Indeed, if there is no such shortcut, then t must be reached starting from a node outside C_δ^n and using only short-range links. But since the length of each short-range link is at most $\sqrt{c \log n}$ and the number of hops is at most n^ϕ, it follows that the total distance travelled by using only local hops is at most $n^\phi \sqrt{c \log n}$ which is, for large enough n, at most n^δ, because $\delta > \phi$. Hence our claim must hold.

Next, we find a bound on the probability of a having at least one shortcut to the disc C_δ^n from the sequence $x_0, x_1, \ldots, x_m$.

Let us start by focusing on the point x_0. We denote the number of nodes in the disc C_δ^n by N_C. This is a Poisson random variable with parameter $\pi n^{2\delta}$ and therefore we have that $N_C < 4n^{2\delta}$ w.h.p. Letting LR_δ be the event that x_0 has a long-range connection to at least one of the N_C nodes in C_δ^n, and noticing that β_n is an upper bound on the probability that there is a shortcut between x_0 and any other node in B_n, we have by the union bound and (6.50) that

$$P(LR_\delta | N_C < 4n^{2\delta}) \leq 16ln^{(4\delta+\alpha-2)/2}. \tag{6.51}$$

As before, this can be viewed as a worst case scenario; conditioning on an event E contained in $\{N_C < 4n^{2\delta}\}$ would lead to the same bound, since β_n is a uniform upper bound.

If x_0 does not have a shortcut, it passes the message to some short-range contact x_1 closer to the target. The only available information about shortcuts at this moment, is that x_0 does not have a shortcut. This, clearly biases the conditional probability for a shortcut from x_1, but according to the observation made above, we do have the same upper bound on the conditional probability of having a shortcut to C_δ^n.

By iteration, we have that the same upper bound holds at at every step in the sequence. Now, by letting LR^ϕ be the event that LR_δ occurs within n^ϕ hops and applying the union bound, we have

$$P(LR^\phi|N_C < 4n^{2\delta}) \leq 16ln^{(2\phi+4\delta+\alpha-2)/2}. \tag{6.52}$$

Notice that (6.52) tends to zero as $n \to \infty$, provided that $\phi < (2-\alpha)/6$ and choosing $\delta > \phi$ small enough such that $2\phi + 4\delta + \alpha - 2 < 0$. Finally, we write

$$P(LR^\phi) = P(LR^\phi|N_C < 4n^{2\delta})P(N_C < 4n^{2\delta})$$
$$+ P(LR^\phi|N_C \geq 4n^{2\delta})P(N_C \geq 4n^{2\delta}), \tag{6.53}$$

and since $P(N_C \geq 4n^{2\delta})$ also tends to zero, we reach a contradiction, and the proof is complete in this case.

Case (iii). The proof of this case is similar to the one of Theorem 6.3.5. The probability that a given node x has a shortcut of length at least r is bounded by the expected number of shortcuts of x that are larger than r, which is bounded by

$$\beta_n \int_r^\infty x^{-\alpha} 2\pi x \, dx \leq \frac{l}{\int_{\sqrt{c\log n}}^{\sqrt{n}/2} x^{-\alpha} 2\pi x \, dx} \int_r^\infty x^{-\alpha} 2\pi x \, dx$$
$$\leq Cr^{2-\alpha} (\log n)^{(\alpha-2)/2}, \tag{6.54}$$

for a uniform constant C and for all n sufficiently large.

Now pick two nodes at random and consider the path that algorithm $\mathcal{A}$ finds between them. Notice that w.h.p. the distance between the two randomly chosen nodes is at least $n^{1/2-\epsilon}$ for any $\epsilon > 0$. It follows that if the path contains at most n^ϕ steps, then there must be one step of length at least $n^{1/2-\epsilon}/n^\phi = n^{1/2-\epsilon-\phi}$. We now compute a bound on the probability of the event A_n that such a step appears in the course of the algorithm, that is in the first n^ϕ steps. By the union bound and (6.54) this is given by,

$$P(A_n) \leq n^\phi C(n^{1/2-\epsilon-\phi})^{2-\alpha}(\log n)^{(\alpha-2)/2}. \tag{6.55}$$

It is easy to see that the exponent of n in the above expression is negative for sufficiently small ϵ and $\gamma < (\alpha-2)/2(\alpha-1)$. It immediately follows that $P(A_n) \to 0$ as n tends to ∞ and hence a route with fewer than n^ϕ hops cannot be found w.h.p. This concludes the proof of the theorem. $\square$

We now describe the last continuum model of this section, which is constructed on the whole plane $\mathbb{R}^2$. This is a simpler model than the previous one mainly because it is based on a tree geometry which has a full independence structure. Thus, the analysis is greatly simplified, as we do not need to worry about biasing the probability law when considering successive steps of the algorithm. The model can be analysed at all distance scales, and naturally leads to the important concept of geometric *scale invariance* in networks, which cannot be appreciated in a discrete setting.

Consider a connection function $g(x) = 1/x^\alpha$, for some $\alpha > 0$ and $x \in \mathbb{R}^+$. Let us construct the model by starting with an arbitrary point $z \in \mathbb{R}^2$. The immediate neighbours of z are given by a non-homogeneous Poisson point process X with density function $\lambda g(|z - y|)$, for some $\lambda > 0$. Similarly, for each Poisson point point $x \in X$ we let its neighbours be given by another Poisson point process, independent of the previous one, and of density function $\lambda g(|x - y|)$. We then iterate in the natural way. Note that each point recieves its 'own' set of neighbours, independently of anyother. Clearly, this is not realistic, but it is designed to fully appreciate the notion of scale invariance which underlies important phenomena in the analysis of random networks.

Let d be the Euclidean distance between a source point $s \in \mathbb{R}^2$ and a target point $t \in \mathbb{R}^2$. For some $\epsilon > 0$, define the ϵ-delivery time of a decentralised algorithm $\mathcal{A}$ as the number of steps required for the message originating at s to reach an ϵ-neighbourhood of t, at each step making the forwarding decision based on the rules of $\mathcal{A}$. Finally, let $\overline{\mathcal{A}}$ be the decentralised algorithm that at each step forwards the message to the local neighbour that is closest in Euclidian distance to the target. We have the following theorem.

Theorem 6.4.2 *The scaling exponent α of the model described above influences the ϵ-delivery time (over a distance d) of a decentralised algorithm as follows:*

(i) *For $\alpha = 2$, there is a constant $c > 0$ such that for any $\epsilon > 0$ and $d > \epsilon$, the expected ϵ-delivery time of the decentralised algorithm $\overline{\mathcal{A}}$ is at most $c(\log d + \log 1/\epsilon)$.*

(ii) *For $\alpha < 2$, there exists a constant $c(\alpha) > 0$ such that for any $\epsilon > 0$, the expected ϵ-delivery time of any decentralised algorithm $\mathcal{A}$ is at least $c(\alpha)(1/\epsilon)^{2-\alpha}$.*

(iii) *For $\alpha > 2$ and any $\epsilon > 0$ and $d > 1$, the expected ϵ-delivery time of any decentralised algorithm $\mathcal{A}$ is at least cd^β, for any $\beta < (\alpha - 2)/(\alpha - 1)$ and some constant $c = c(\alpha, \beta) > 0$.*

Notice that since the model above is continuous and defined on the whole plane, it allows us to appreciate all distance scales. Essentially, the theorem says that for $\alpha = 2$ it is possible to approach the target at any distance scale in a logarithmic number of steps, steadily improving at each step. On the other hand, when $\alpha < 2$ a decentralised algorithm starts off quickly, but then slows down as it approaches the target, having trouble making the last small steps. For $\alpha > 2$, the situation is reversed, as the performance bottleneck is not near the target, but is at large distances $d \gg \epsilon$.

Proof of Theorem 6.4.2. Case (i). Let $V \subset \mathbb{R}^2$ be a bounded set, not containing the origin, over which the Lebesgue integral of g can be defined. The (random) number of acquaintances of O in V has a Poisson distribution with mean $\int_V g(x)dx$. A simple substitution shows that,

$$\int_{aV} g(x)dx = a^{2-\alpha} \int_V g(x)dx, \qquad (6.56)$$

where aV is the set $\{av; v \in V\}$. From this it follows that $\alpha = 2$ is a special case, since in this case $\int_{aV} g(x)dx$ is independent of a. The interpretation of this is that for the case

$\alpha = 2$, the model has no natural scale; changing the unit of length does not make any difference.

We now compute the probability that at any step of the algorithm an intermediate node has a neighbour at a distance from the target that is less than half the distance between the target and the intermediate node, and show that this probability is positive and independent of distance.

We refer to Figure 6.7. Let $\overline{OT} = r$ be the distance to the target. The (random) number of neighbours that are at a distance less than $r/2$ from the target t has a Poisson distribution with mean

$$\mu = \lambda \int_{-\pi/6}^{\pi/6} \int_{A_r(\theta)}^{B_r(\theta)} g(x) x \, dx \, d\theta, \tag{6.57}$$

which is positive and, since $\alpha = 2$, also independent of r. It follows that there is always a positive probability $\tau = 1 - e^{-\mu}$, independent of r, that point O has a neighbour inside the circle depicted in Figure 6.7, i.e., closer to t than O by at least half the distance between t and O. Hence, algorithm $\overline{\mathcal{A}}$, forwarding the message to the node closest to the target, can reduce the distance to the target by a factor of at least $1/2$ with uniform positive probability at each step. Whenever this occurs we say that the algorithm has taken a *successful* step. We have seen that a successful step has uniform positive probability, we now show that a step that simply decreases the distance to the target has probability one. The number of points that are closer than r to the target is again Poisson distributed, with mean given by the integral of λg over the disc of radius r centred at t. It is easy to see that this integral diverges, and hence this number is infinite with probability one. It follows that the probability of *decreasing* the distance to the target has probability one. Hence, even when a step of the algorithm is not successful, it will not increase the distance to the target. It follows that at most a total number of n successful steps are needed to reach an ϵ-neighbourhood of t, starting at a distance $d > \epsilon$, where

$$\left(\frac{1}{2}\right)^n d < \epsilon \Leftrightarrow n < \frac{\log d + \log 1/\epsilon}{\log 2}. \tag{6.58}$$

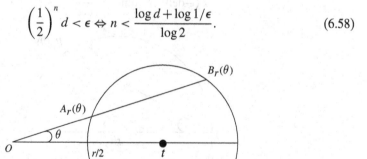

Fig. 6.7   Decreasing the distance to the target by a factor 1/2.

The expected waiting time for the nth successful step is n/τ, and therefore our bound on the expected ϵ-delivery time is

$$E(\epsilon\text{-delivery time}) < \frac{\log d + \log 1/\epsilon}{\tau \log 2}, \tag{6.59}$$

which concludes the proof in this case.

Case (ii). We consider a generic step of an algorithm, where the message is at point O, at distance $r \geq \epsilon$ from the target and start by computing the number of neighbours of point O that are closer to the target. We refer to Figure 6.8. The number of such points has a Poisson distribution, and since $\alpha < 2$ it has a finite mean

$$\mu(r, \alpha) = \lambda \int_{-\pi/2}^{\pi/2} \int_0^{B_r(\theta)} g(r) r \, dr \, d\theta$$

$$= \lambda \int_{-\pi/2}^{\pi/2} \int_0^{rB_1(\theta)} \frac{1}{r^{\alpha-1}} dr \, d\theta$$

$$= \frac{\lambda}{2-\alpha} r^{2-\alpha} \int_{-\pi/2}^{\pi/2} B_1(\theta)^{2-\alpha} d\theta$$

$$= c(\alpha) r^{2-\alpha}. \tag{6.60}$$

Let an *improving* step of any decentralised algorithm be one that forwards the message to a neighbour that is closer to the target than is O. The above computation shows that when the message is at distance ϵ from the source, the probability for an improving step is bounded above by $c(\alpha)\epsilon^{2-\alpha}$. When the distance to the target is larger than ϵ, the probability to enter the ϵ-neighbourhood is easily seen to be smaller than this probability, since the density of the Poisson processes decreases with distance. Hence, at *any* step in the algorithm the probability of an ϵ-delivery is at most $c(\alpha)\epsilon^{2-\alpha}$. It follows that the expected number of steps required to enter an ϵ-neighbourhood of the target is at least

$$E(\epsilon\text{-delivery time}) \geq \frac{1}{c(\alpha)\epsilon^{2-\alpha}}. \tag{6.61}$$

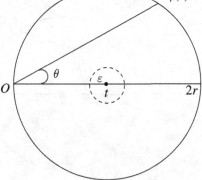

Fig. 6.8   Getting closer to the target.

Case (iii). Consider the collection of neighbours of a given Poisson point, and denote by D the distance to the neighbour furthest away. We find that

$$P(D > r) = 2\pi\lambda \int_r^\infty x^{-\alpha} x \, dx$$

$$= \frac{c}{\alpha - 2} r^{2-\alpha}, \tag{6.62}$$

for some constant c. This quantity tends to zero as $r \to \infty$, since $\alpha > 2$.

We next estimate the probability that starting at distance $d > 1$, an ϵ-delivery can take place in at most d^β steps, for some $\beta > 0$. Delivery in at most d^β steps implies that in one of the first d^β steps of the algorithm, there must be at least one step of size at least $d^{1-\beta}$. According to the computation above, the probability that this happens is at most

$$\frac{c}{\alpha - 2} d^\beta d^{(1-\beta)(2-\alpha)} = d^{2-\alpha-\beta+\alpha\beta}. \tag{6.63}$$

Writing X_d for the delivery time starting at distance d, it follows that

$$P(X_d \geq d^\beta) \geq 1 - \frac{c}{\alpha - 2} d^{2-\alpha-\beta+\alpha\beta} \tag{6.64}$$

and therefore that

$$E(X_d) \geq d^\beta \left(1 - \frac{c}{\alpha - 2} d^{2-\alpha-\beta+\alpha\beta} \right). \tag{6.65}$$

Whenever $2 - \alpha - \beta + \alpha\beta < 0$ or, equivalently,

$$\beta < \frac{\alpha - 2}{\alpha - 1}, \tag{6.66}$$

this expression is, for some constant c, at least cd^β. The result now follows. $\square$

6.5 The role of scale invariance in networks

The long-range percolation models we have described exhibit a transition point for efficient navigation at a critical scaling exponent $\alpha = 2$. In this section we want to make some observations on this peculiar property, introducing the concept of *scale invariance* in networks. Since the term scale invariant is used in many scientific contexts, it is worth spending a few words on it, to clarify precisely what it means in our specific context. We shall argue that scale invariance plays an important role in random networks, and it can be used to provide guidelines for both the analysis and design of real networks.

Scale invariance refers to objects or laws that do not change when the units of measure are multiplied by a common factor. This is often the case for statistical physics models at criticality. The situation is best described by focusing on bond percolation on the square lattice. Consider a box B_n of size $n \times n$ and a given value of the parameter $p \neq p_c$. If the value of p is changed by a small amount, one expects this to change the state of only a few bonds and not to greatly affect the connectivity properties of the system. However, if p is near p_c and the box B_n is large, then changing p by a small amount may have a dramatic effect on connectivity over large distances. A typical observation is then that *at* the critical point, fluctuations occur at all scale lengths, and thus one should look for a scale-invariant theory to describe the behaviour of the system.

Indeed, at the critical point the appearance of the system is essentially not influenced by the scale at which we observe it. For instance, we recall from Chapter 4 that at criticality, the probability of finding a crossing path in the box B_n is a constant equal to $1/2$, and hence *independent* of the box size. On the other hand, we have also seen that above or below criticality the appearance of the system is very much influenced by the scale at which we observe it. For example, above criticality, as one looks over larger and larger boxes, crossings paths, that could not be observed inside smaller boxes, eventually appear. Similarly, below criticality, one might be able to observe some crossings in a small box, but as the size of the box increases all crossings tend to disappear.

Naturally, the characteristic scale length at which one observes the appearance or disappearance of the crossing paths depends on the value of p, and as $p \to p_c$ this characteristic scale diverges. The conclusion is that while above and below criticality there is a natural characteristic scale at which to observe the system, at criticality there is not.

Now, the critical exponent $\alpha = 2$ that we have observed in the context of navigation of random networks is also related to the scale invariance phenomenon described above. Recall from (6.56) that the exponent $\alpha = 2$, the value of which is dictated by the dimension of the space, is the one for which changing the units of length in the model does not make any difference in its geometric structure. For this exponent, each Poisson point has the same distribution of connections at all scale lengths, which turns out to be essential to efficiently reach an arbitrary destination on the plane. We have also seen that the critical exponent $\alpha = 2$ arises in at least two other models of long-range percolation which are, at the microscopic level, quite different from each other. In other words, we have seen three long-range percolation models which belong to a class for which $\alpha = 2$ is the *universal* exponent describing the scale-invariant phenomenon.

The observation that different models may belong to the same universality class, and hence have the same critical exponents is also typical in statistical physics. In a scale-invariant setting, we expect similar phenomena to be observed at all scale lengths, and hence the quantities that describe them to display the same functional form, regardless of the microscopic structure of the underlying graph; and this was indeed the case for our navigation model. In statistical physics, characteristic functions of different percolation models are believed to be described near the critical point by power laws with the same critical exponents.

It is, finally, tempting to make a heuristic leap and apply the universality principle stating that in dimension two, $\alpha = 2$, being independent of the local structure of the model, is the correct scaling to use in the design of communication networks to facilitate multi-hop routing. Of course, this is far from being a rigorous statement.

6.6 Historical notes and further reading

Although long-range percolation models have been considered in mathematics for quite some time, an important paper that drew renewed attention to the drastic reduction of the diameter of a network when a few long-range connections are added at random, is the one by Watts and Strogatz (1998), who observed the phenomenon by computer simulations. Bollóbas and Chung (1988) showed earlier similar results rigorously, by adding a random matching to the nodes of a cycle. Watts and Strogatz's paper, however, played a critical role in sparking much of the activity on modelling real world networks via random graphs constructed using simple rules.

A proof of Theorem 6.2.2 on the chemical distance can be found in Antal and Pisztora (1996), while the navigation Theorem 6.2.3 is by Angel, Benjamini *et al.* (2005), who also prove a more general result in any dimension. Theorems 6.3.2 and 6.3.3 are by Coppersmith, Gamarnik and Sviridenko (2002). Of course, there are many other models in the literature that exhibit small diameters when the parameters are chosen appropriately. Yukich (2006), for example, has considered a geometric model of 'ultra small' random networks, in which the graph distance between x and y scales as $\log\log|x - y|$.

The algorithmic technique using the crossbar construction to describe navigation along the highway follows an idea presented in a different context by Kaklamanis, Karlin, *et al.* (1990). An important paper that drew renewed attention to the difference between the existence of paths in random graphs and their algorithmic discovery is the one by Kleinberg (2000), who was inspired by the so-called 'small world phenomenon': a mixture of anecdotal evidence and experimental results suggesting that people, using only local information, are very effective at finding short paths in a network of social contacts. Theorem 6.3.6 presents a slight variation of Kleinberg's original proof. The continuum versions of the result, namely Theorems 6.4.1 and 6.4.2, are by Ganesh and Draief (2006), and Franceschetti and Meester (2006b), respectively. In the former paper, a stronger version, without ϵ in the exponent, is announced. It is not hard to prove this, using a Chernoff bound. Similarly, a.s. statements of Theorem 6.4.2 can also be obtained.

Some different algorithmic issues also related to percolation and coverings can be found in Booth *et al.* (2003).

Scale invariance, critical exponents, and universality have a long and rich history in the study of disordered systems. Physicists invented the renormalisation group to explain these phenomena that were experimentally observed, but this method has not been made mathematically rigorous for percolation or other models of random networks. In the last few years, key advancements by Lawler, Schramm, Smirnov, and Werner, have proved power laws for site percolation on the triangular lattice approaching criticality, and confirmed many values of critical exponents predicted by physicists. The proofs are based on Schramm's invention of stochastic Loewner's evolutions, on Kesten's scaling relations, and on Smirnov's proof of the existence of conformal invariance properties of certain crossing probabilities. For an account of these works, we refer the reader to Smirnov (2001), Smirnov and Werner (2001), and the references therein; and also to Camia and Newman (2007), who worked out Smirnov's proof in great detail. Proving the existence of power laws and universality in models other than the triangular lattice remains one of the main open problems in mathematical physics today.

Exercises

6.1 Prove Theorem 6.3.5 employing a navigation length definition that considers the worst case scenario of the number of steps required to connect *any* two vertices of the random network.

6.2 Identify where in the proof of Theorem 6.4.1 the assumption of having a torus rather than a square has been used.

6.3 Prove in the context of Theorem 6.4.1 that if one defines the algorithm at each node to simply forward the message to the neighbour y closest to the target t, then the conditional law of the point process in the circle centred at the target and of radius $|t - y|$ is not Poisson.

6.4 Show that when X_p has a geometric distribution with parameter p, then as $p \to 0$, $P(pX_p \le x)$ converges to $P(Y \le x)$ where Y has an exponential distribution. Use this to show that in Case (i) of Theorem 6.4.1, it is the case that $P(S(x) > K \log n) \to 0$ as $K \to \infty$, *uniformly* in n.

6.5 Finish the proof of Theorem 6.4.1, Case (i).

6.6 Prove that the bounds on the diameter in Theorem 6.4.2 also hold with high probability.

Appendix

In this appendix we collect a number of technical items that are used in the text, but which we did not want to work out in the main text in order to keep the flow going.

A.1 Landau's order notation

We often make use of the standard so-called 'order notation', which is used to simplify the appearance of formulas by 'hiding' the uninteresting terms. In the following x_0 can be $\pm\infty$. When we write

$$f(x) = O\left(g(x)\right) \text{ as } x \to x_0, \tag{A.1}$$

we mean that

$$\limsup_{x \to x_0} \frac{f(x)}{g(x)} < \infty. \tag{A.2}$$

When we write

$$f(x) = o\left(g(x)\right) \text{ as } x \to x_0, \tag{A.3}$$

we mean that

$$\lim_{x \to x_0} \frac{f(x)}{g(x)} = 0. \tag{A.4}$$

A.2 Stirling's formula

Stirling's formula can be found in just about any introductory textbook in calculus or analysis. It determines the rate of growth of $n!$ as $n \to \infty$. It reads as follows:

$$\lim_{n \to \infty} \frac{n!}{n^{(n+1/2)} e^{-n}} = \sqrt{2\pi}. \tag{A.5}$$

A.3 Ergodicity and the ergodic theorem

The ergodic theorem can be viewed as a generalisation of the classical strong law of large numbers (SLLN). Here we only present a very informal discussion of the ergodic theorem. For more details, examples and proofs, see the book by Meester and Roy (1996).

Informally, the classical SLLN states that the average of many independent and identically distributed random variables is close to the common expectation. More precisely, if..., $X_{-1}, X_0, X_1, X_2, \ldots$ are i.i.d. random variables with common expectation μ, then the average

$$\frac{1}{2n+1} \sum_{i=-n}^{n} X_i \qquad\qquad (A.6)$$

converges to μ with probability one, as $n \to \infty$.

It turns out that this result is true in many circumstances where the independence assumption is replaced by the much weaker assumption of *stationarity*. In this context, we say that the sequence of random variables $\ldots, X_{-1}, X_0, X_1, X_2, \ldots$ is *stationary* if the distribution of the random vector

$$(X_k, X_{k+1}, \ldots, X_{k+m}) \qquad\qquad (A.7)$$

does not depend on the starting index k. In particular this implies (by taking $m = 0$) that all the X_i have the same distribution; they need no longer be independent though.

We would like to have a SLLN in the context of stationary sequences, but a little reflection shows that this can not be the case in general. Indeed, if for example we let all the X_i take the same value zero (simultaneously) with probability $1/2$, and the value one (simultaneously) with probability $1/2$, then the average of $X_{-n}, \ldots, X_n$ converges (in fact, is equal) to one with probability $1/2$ and converges to zero also with probability $1/2$. At the same time, the common expectation of the X_i is $1/2$ and therefore the SLLN does not hold in this case.

This example shows that if a stationary sequence is the 'combination' of two other stationary sequences, then the SLLN need not hold. It turns out that *not* being such a 'combination' is precisely the condition which makes the SLLN true. Indeed, informally, the *ergodic theorem* states that if a sequence can *not* be written as any such combination, then the SLLN does hold, and the average of the random variables does converge to the common expectation, with probability 1. The combinations that we talk about here need not be combinations of just two or even only a countable number of stationary processes. For example, one can construct a combination by first drawing a uniform $(0, 1)$ random variable Y, and if Y takes the value y, let all the X_i be equal to y. Then the X_i process is the combination of an uncountable number of other stationary sequences.

This discussion – of course – begs the question as to when a sequence of random variables can not be written as a combination of two other stationary sequences. This is not a trivial matter and it is out of the scope of this book. It suffices to say that in all cases where we use the ergodic theorem in this book, this assumption is met. When this assumption is met, we say that the sequence of the X_i is *ergodic*.

In fact, there is also a two-dimensional version of the ergodic theorem which is even more important to us. In the two-dimensional case we do not talk about a sequence of random variables, but about an *array* of such random variables, indexed by (i, j), for integers i and j. An example of such an array in the context of this book is the following. Let $X_{i,j}$ be the number of isolated nodes in the square $S_{i,j} = [i, i+1] \times [j, j+1]$ in a boolean model of densitiy $\lambda > 0$. The $X_{i,j}$ are not independent, but they are stationary in the obvious two-dimensional sense. The ergodic theorem in this case now tells us that the average number of isolated points in all squares $S_{i,j}$, with $-n \leq i, j < n$ converges, as $n \to \infty$, to the expected number of such isolated nodes in the unit square. This is a typical application of the ergodic theorem in this book.

Finally, we mention a property of ergodic sequences and arrays, which is sometimes even used as a definition. When we have such an ergodic sequence or array, any event which is invariant under translations, will have probability either zero or one. For example, the event that a certain percolation model has an infinite component, is a translation-invariant event, since when we shift all vertices simultaneously, the infinite component will also shift, but remains infinite. The event that the origin is in an infinite component is *not* invariant under such translations, and indeed the probability of this event need not be restricted to zero or one.

A.4 Deviations from the mean

A.4.1 Markov's inequality

For a random variable X such that $P(X \geq 0) = 1$, and for all $n \geq 0$, we have

$$P(X \geq x) \leq \frac{E(X^n)}{x^n}. \tag{A.8}$$

Another version of the above inequality reads as follows: for all $s \geq 0$ we have

$$P(X \geq x) \leq e^{-sx} E(e^{sX}). \tag{A.9}$$

Proof

$$
\begin{aligned}
E(X^n) &= \int x^n dP_X \\
&= \int_{X<x} x^n dP_X + \int_{X \geq x} x^n dP_X \\
&\geq \int_{X \geq x} x^n dP_X \\
&\geq x^n P(X \geq x).
\end{aligned}
\tag{A.10}
$$

The other version follows along the same lines. $\square$

A.4.2 Chebyshev's inequality

Let $\mu = E(X)$; this inequality is obtained from Markov's inequality by substituting $|X - \mu|$ for X and taking $n = 2$:

$$P(|X - \mu| \geq \mu) \leq \frac{Var(X)}{\mu^2}. \tag{A.11}$$

A.4.3 Chernoff's bounds for a Poisson random variable

For a Poisson random variable X with parameter λ, we have

$$P(X \geq x) \leq \frac{e^{-\lambda}(e\lambda)^x}{x^x} \text{ for } x > \lambda, \tag{A.12}$$

$$P(X \leq x) \leq \frac{e^{-\lambda}(e\lambda)^x}{x^x} \text{ for } x < \lambda. \tag{A.13}$$

Proof

$$E(e^{sX}) = \sum_{k=0}^{\infty} \frac{e^{-\lambda}\lambda^k}{k!} e^{sk}$$

$$= e^{\lambda(e^s-1)} \sum_{k=0}^{\infty} \frac{e^{-\lambda e^s}(\lambda e^s)^k}{k!}$$

$$= e^{\lambda(e^s-1)}. \tag{A.14}$$

For any $s > 0$ and $x > \lambda$ applying Markov's inequality we have

$$P(X \geq x) = P(e^{sX} > e^{sx}) \leq \frac{E(e^{sX})}{e^{sx}} = e^{\lambda(e^s-1)-sx}. \tag{A.15}$$

Letting $s = \log(x/\lambda) > 0$ we finally obtain

$$P(X \geq x) \leq e^{x-\lambda-x\log(x/\lambda)} = \frac{e^{-\lambda}(e\lambda)^x}{x^x}. \tag{A.16}$$

The lower tail bound follows from similar computations. $\qquad\square$

A.5 The Cauchy–Schwarz inequality

For two random variables X, Y defined on the same sample space, with $E(X) < \infty$, $E(Y) < \infty$, we have

$$E^2(XY) \leq E(X^2)E(Y^2). \tag{A.17}$$

Proof Let a be a real number and let $Z = aX - Y$. We have that

$$0 \leq E(Z^2) = a^2 E(X^2) - 2aE(XY) + E(Y^2). \tag{A.18}$$

This can be seen as a quadratic inequality in the variable a. It follows that the discriminant is non-positive. That is, we have

$$(2E(XY))^2 - 4E(X^2)E(Y^2) \leq 0, \tag{A.19}$$

which gives the desired result. $\qquad\square$

A.6 The singular value decomposition

For any $m \times n$ real (or complex) matrix M, there exists a factorisation of the form

$$M = USV^*, \tag{A.20}$$

where U is an $m \times m$ unitary matrix, S is an $m \times n$ matrix with non-negative numbers on the diagonal and zeros off the diagonal, and V^* is the conjugate transpose of V, which is an $n \times n$ unitary matrix. Such a factorisation is called a singular value decomposition of M. The elements of S are called the singular values, and the columns of U and V are the left and right singular vectors of the corresponding singular values.

The singular value decomposition can be applied to any $m \times n$ matrix. The eigenvalue decomposition, on the other hand, can only be applied to certain classes of square matrices. Nevertheless, the two decompositions are related. In the special case that M is Hermitian, the singular values and the singular vectors coincide with the eigenvalues and eigenvectors of M. The following relations hold:

$$M^*M = V(S^*S)V^*$$

$$MM^* = U(SS^*)U^*. \tag{A.21}$$

The right-hand side of above relations describe the eigenvalue decompositions of the left-hand sides. Consequently, the squares of the singular values of M are equal to the eigenvalues of MM^* or M^*M. Furthermore, the left singular vectors of U are the eigenvectors of MM^* and the right singular vectors are the eigenvectors of M^*M.

References

Aizenman, M., J. Chayes, L. Chayes, J. Frölich, L. Russo (1983). On a sharp transition from area law to perimeter law in a system of random surfaces. *Communications in Mathematical Physics* **92**, 19–69.

Aizenman, M., H. Kesten, C. Newman (1987). Uniqueness of the infinite cluster and continuity of connectivity functions for short- and log-range percolation. *Communications in Mathematical Physics* **111**, 505–32.

Alexander, K. (1991). Finite clusters in high density continuum percolation: compression and sphericality. *Probability Theory and Related Fields* **97**, 35–63.

Angel, O., I. Benjamini, E. Ofek, U. Wieder (2005). Routing complexity of faulty networks. *Proceedings of the Twenty-Fourth Annual ACM Symposium on Principles of Distributed Computing*, 209–17.

Antal, P., A. Pisztora (1996). On the chemical distance for supercritical Bernoulli percolation. *Annals of Probability* **24**(2), 1036–48.

Arratia, R., L. Goldstein, L. Gordon (1989). Two moments suffice for Poisson approximations: the Chen–Stein method. *Annals of Probability* **17**(1), 9–25.

Balister, P., B. Bollobás, A. Sarkar, M. Walters (2005). Connectivity of random k-nearest neighbour graphs. *Advances in Applied Probability* **37**(1), 1–24.

Balister, P., B. Bollobás, M. Walters (2004). Continuum percolation with steps in an annulus. *Annals of Applied Probabilty* **14**(4), 1869–79.

Barbour, A. D., L. Holst, S. Janson (1992). *Poisson Approximation*. Oxford: Clarendon Press.

Barlow, R., F. Proschan (1965). *Mathematical Theory of Reliability*. New York: John Wiley & Sons.

Berg, van den, R., H. Kesten (1985). Inequalities with applications to percolation and reliability. *Journal of Applied Probability* **22**, 556–69.

Bollobás, B. (2001). *Random Graphs*. Cambridge: Cambridge University Press.

Bollobás, B., F. Chung (1988). The diameter of a cycle plus a random matching. *SIAM Journal of Discrete Mathematics* **1**(3), 328–33.

Bollobás, B., O. Riordan (2006). *Percolation*. Cambridge: Cambridge University Press.

Booth, L., J. Bruck, M. Franceschetti, R. Meester (2003). Covering algorithms, continuum percolation, and the geometry of wireless networks. *Annals of Applied Probability* **13**(2), 722–31.

Broadbent, S. R., J. M. Hammersley (1957). Percolation processes I. Crystals and mazes. *Proceedings of the Cambridge Philosophical Society*, **53**, 629–41.

Brug, van de, T. (2003). The Poisson random connection model: construction, central limit theorem and asymptotic connectivity. Master thesis, Vrije Universiteit Amsterdam.

Burton, R., M. Keane (1989). Density and uniqueness in percolation. *Communications in Mathematical Physics* **121**, 501–5.

Camia, F., C. Newman (2007). Critical percolation exploration path and SLE_6: a proof of convergence. *Probability Theory and Related Fields*. In press.

Chen, L. H. Y. (1975). Poisson approximation for dependent trials. *Annals of Probability* 3, 534–45.

Coppersmith, D., D. Gamarnik, M. Sviridenko (2002). The diameter of a long range percolation graph. *Random Structures and Algorithms*, 21(1), 1–13.

Cover, T., J. Thomas (2006). *Elements of Information Theory*. New York: John Wiley & Sons.

Cox, J., R. Durrett (1988). Limit theorems for spread out of epidemics and forest fires. *Stochastic Processes and Their Applications* 30, 171–91.

Daley, D., D. Vere-Jones (1988). *An Introduction to the Theory of Point Processes*. Berlin: Springer Verlag.

Dousse, O. (2005). Asymptotic properties of wireless multi-hop networks. Ph.D. thesis, École Polytechnique Fédérale de Lausanne.

Dousse, O., F. Baccelli, P. Thiran (2005). Impact of interferences on connectivity in ad-hoc networks. *IEEE/ACM Transactions on Networking*, 13(2), 425–36.

Dousse, O., M. Franceschetti, N. Macris, R. Meester, P. Thiran (2006). Percolation in the signal to interference ratio graph. *Journal of Applied Probability* 43(2), 552–62.

Dousse, O., M. Franceschetti, P. Thiran (2006). On the throughput scaling of wireless relay networks. *IEEE Transactions on Information Theory* 52(6), 2756–61.

Durrett, R. (2007). *Random Graph Dynamics*. Cambridge: Cambridge University Press.

Erdös, P., A. Rényi (1959). On random graphs. *Publicationes Mathematicae Debrecen* 6, 290–7.

Erdös, P., A. Rényi (1960). On the evolution of random graphs. *Tudományos Akadémia Matematikai Kutató Intézetének Közleményei* 5, 17–71.

Erdös, P., A. Rényi (1961). On the strength of connectedness of a random graph. *Acta Mathematica Academiae Scientiarum Hungaricum* 12, 261–7.

Fortuin, C., C. Kasteleyn, J. Ginibre (1971). Correlation inequalities on some partially ordered sets. *Communications in Mathematical Physics* 22, 89–103.

Franceschetti, M. (2007). A note on Lévéque and Telatar's upper bound on the capacity of wireless ad-hoc networks. *IEEE Transactions on Information Theory* 53(9), 3207–11.

Franceschetti, M., L. Booth, M. Cook, R. Meester, J. Bruck (2005). Continuum percolation with unreliable and spread-out connections. *Journal of Statistical Physics* 118(3–4), 719–31.

Franceschetti, M., O. Dousse, D. Tse, P. Thiran (2007). Closing the gap in the capacity of wireless networks via percolation theory. *IEEE Transactions on Information Theory* 53(3), 1009–18.

Franceschetti, M., R. Meester (2006a). Critical node lifetimes in random networks via the Chen-Stein method. *IEEE Transactions on Information Theory* 52(6), 2831–7.

Franceschetti, M., R. Meester (2006b). Navigation in small world networks, a continuum, scale-free model. *Journal of Applied Probability* 43(4), 1173–80.

Friedgut, E., G. Kalai (1996). Every monotone graph property has a sharp threshold. *Proceedings of the American Mathematical Society* 124, 2993–3002.

Ganesh, A., M. Draief (2006). Efficient routing in Poisson small-world networks. *Journal of Applied Probability* 43(3), 678–86.

Gilbert, E. N. (1961). Random plane networks. *Journal of SIAM* 9, 533–43.

Goel, A., S. Rai, B. Krishnamachari (2005). Monotone properties of random geometric graphs have sharp thresholds. *Annals of Applied Probability* 15(4), 2535–52.

Gonzáles-Barrios, J. M., A. J. Quiroz (2003). A clustering procedure based on the comparison between the k nearest neighbors graph and the minimal spanning tree. *Statistics and Probability Letters* 62, 23–34.

Grimmett, G. (1999). *Percolation*. Berlin: Springer Verlag.

Grimmett, G., A. Stacey (1998). Critical probabilities for site and bond percolation models. *Annals of Probability* **26**(4), 1788–1812.

Grimmett, G., D. Stirzaker (1992). *Probability and Random Processes.* Oxford: Oxford University Press.

Gupta, P., P. R. Kumar (1998). Critical power for asymptotic connectivity in wireless networks. In *Stochastic Analysis, Control, Optimization and Applications: A Volume in Honor of W. H. Fleming,* eds. W. M. McEneaney, G. Yin, Q. Zhang. Boston: Birkhäuser, pp. 547–67.

Gupta, P., P. R. Kumar (2000). The capacity of wireless networks. *IEEE Transactions on Information Theory* **46**(2), 388–404.

Häggström, O., R. Meester (1996). Nearest neighbor and hard sphere models in continuum percolation. *Random Structures and Algorithms* **9**(3), 295–315.

Harris, T. (1960). A lower bound on the critical probability of a certain percolation process. *Proceedings of the Cambridge Philosophical Society* **56**, 13–20.

Harris, T. (1963). *The Theory of Branching Processes.* New York: Dover.

Kaklamanis, C., A. Karlin, F. Leighton, *et al.* (1990). Asymptotically tight bounds for computing with faulty arrays of processors. *Proceedings of the 31st Annual Symposium on the Foundations of Computer Science,* 285–96.

Kesten, H. (1980). The critical probability of bond percolation on the square lattice equals 1/2. *Communications in Mathematical Physics* **74**, 41–59.

Kesten, H. (1982). *Percolation Theory for Mathematicians.* Boston: Birkhäuser.

Kingman, J. (1992). *Poisson Processes.* Oxford: Clarendon Press.

Kleinberg, J. (2000). The small-world phenomenon: an algorithmic perspective. *Proceedings of the 32nd ACM Symposium on the Theory of Computing,* 163–70.

Lévêque, O., E. Telatar (2005). Information theoretic upper bounds on the capacity of large extended ad hoc wireless networks. *IEEE Transactions on Information Theory* **51**(3), 858–65.

Liggett, T. M., R. H. Schonmann, A. M. Stacey (1997). Domination by product measures. *Annals of Probability* **25**(1), 71–95.

McEliece, R. J. (2004). *The Theory of Information and Coding.* Cambridge: Cambridge University Press.

Meester, R., T. van de Brug (2004). On central limit theorems in the random connection model. *Physica A* **332**, 263–78.

Meester, R., M. D. Penrose, A. Sarkar (1997). The random connection model in high dimensions. *Statistics and Probability Letters* **35**, 145–53.

Meester, R., R. Roy (1994). Uniqueness of unbounded and vacant components in boolean models. *Advances in Applied Probability* **4**(3), 933–51.

Meester, R., R. Roy (1996). *Continuum Percolation.* Cambridge: Cambridge University Press.

Moore, E., C. Shannon (1956). Reliable circuits using less reliable relays I, II. *Journal of the Franklin Institute* **262**, 191–208, 281–97.

Nyquist, H. (1924). Certain factors affecting telegraph speed. *Bell Systems Technical Journal* **3**, 324.

Özgür, A., O. Lévêque, D. Tse (2007). Hierarchical cooperation achieves optimal capacity scaling in ad hoc networks. *IEEE Transactions on Information Theory,* **53**(10), 3549–72.

Peierls, R. (1936). On Ising's model of ferromagnetism. *Proceedings of the Cambridge Philosophical Society* **36**, 477–81.

Penrose, M. D. (1991). On a continuum percolation model. *Advances in Applied Probability* **23**(3), 536–56.

Penrose, M. D. (1993). On the spread-out limit for bond and continuum percolation. *Annals of Applied Probability* **3**(1), 253–76.

Penrose, M. D. (1997). The longest edge of the random minimal spanning tree. *Annals of Applied Probability* **7**(2), 340–61.

Penrose, M. D. (2003). *Random Geometric Graphs*. Oxford: Oxford University Press.

Penrose, M. D., A. Pisztora (1996). Large deviations for discrete and continuous percolation. *Advances in Applied Probability* **28**(1), 29–52.

Roy, R., A. Sarkar (2003). High density asymptotics of the Poisson random connection model. *Physica A* **318**, 230–42.

Russo, L. (1978). A note on percolation. *Zeitschrift für Wahrscheinlichkeitstheorie Verwandte Gebiete* **43**, 39–48.

Russo, L. (1981). On the critical probabilities. *Zeitschrift für Wahrscheinlichkeitstheorie Verwandte Gebiete* **56**, 229–37.

Russo, L. (1982). An approximate zero-one law. *Zeitschrift für Wahrscheinlichkeitstheorie Verwandte Gebiete* **61**, 129–39.

Seymour, P. D., D. J. A. Welsh (1978). Percolation probabilities on the square lattice. In *Advances in Graph Theory*. Vol. 3 of *Annals of Discrete Mathematics*, ed. B. Bollobás. Amsterdam: North Holland, pp. 227–45.

Shannon, C. (1948). A mathematical theory of communication. *Bell Systems Technical Journal* **27**, 379–423, 623–56. *Reprinted as: The Mathematical Theory of Communication*. Champaign: University of Illinois Press.

Smirnov, S. (2001). Critical percolation in the plane: conformal invariance, Cardy's formula, scaling limits. *Les Comptes Rendus de l'Académie des Sciences, Série I, Mathematique* **333**(3), 239–44.

Smirnov, S., W. Werner (2001). Critical exponents for two-dimensional percolation. *Mathematical Research Letters* **8**, 729–44.

Stein, C. (1978). Asymptotic evaluation of the number of latin rectangles. *Journal of Combinatorial Theory A* **25**, 38–49.

Talagrand, M. (1994). On Russo's approximate zero-one law. *Annals of Probability* **22**(3), 1576–87.

Telatar, E. (1999). Capacity of multi-antenna Gaussian channels. *European Transactions on Telecommunications* **10**(6), 585–95.

Watson, H., F. Galton (1874). On the probability of extinction of families. *Journal of the Anthropology Institute of Great Britain and Ireland* **4**, 138–44.

Watts, D. J., S. H. Strogatz (1998). Collective dynamics of small-world networks. *Nature* **393**, 440–2.

Wilson, R. J. (1979). *Introduction to Graph Theory*. London: Longman.

Xie, L. L., P. R. Kumar (2004). A network information theory for wireless communication: scaling laws and optimal operation. *IEEE Transactions on Information Theory* **50**(5), 748–67.

Xue, F., P. R. Kumar (2004). The number of neighbors needed for connectivity of wireless networks. *Wireless Networks* **10**, 169–81.

Xue, F., L. L. Xie, P. R. Kumar (2005). The transport capacity of wireless networks over fading channels. *IEEE Transactions on Information Theory*, **51**(3), 834–47.

Yukich, J. E. (2006). Ultra-small scale-free geometric networks. *Journal of Applied Probability* **43**(3), 665–77.

Zhang, Y. (1988). Proof of Lemma 11.12. In G. Grimmett (1999), *Percolation*. Berlin: Springer Verlag.

Zong, C. (1998). The kissing number of convex bodies. A brief survey. *Bulletin of the London Mathematical Society* **30**(1), 1–10.

Index

Printed in the United States
By Bookmasters